T0248418

Badvertising

'Brilliant work ... if you thought your brain was being gently warmed by the advertising industry, read this book and you'll realise it's being fried. I couldn't believe just how effective and how underhand some of the tactics being used in advertising are – *Badvertising* shows how we are all prisoners, but it also passes us the keys to our cells. This book was a watershed moment for me. Since it can't have an advertising campaign, we all need to tell our friends about it.'

—Jeremy Vine, broadcaster, journalist,
host of BBC Radio 2's *Jeremy Vine* show

'A hugely timely and important book that grapples with advertising's role in enabling climate crimes – and sets out why and how we need to stop the industry's complicity in its tracks, for the sake of a liveable future.'

—Caroline Lucas MP

'Why do we allow adverts that actively promote our own destruction? Halting climate catastrophe is hard enough without ads selling things that pollute more. It's extraordinary the simple case for tobacco-style bans hasn't been made so clearly before. With *Badvertising*, Simms and Murray have done the world an urgent favour. Funny and readable, [it] will make us all see the world and the advertising we are immersed in 24/7 in a very different way.'

—Dr Chris van Tulleken, infectious diseases doctor,
broadcaster and author of *Ultra-Processed People*

'Simms and Murray are right to lay so much blame at the door of the advertising industry ... [they] are clear-headed guides through the fog. Learn the history, be enraged at the tactics, and join the struggle for a less polluted public sphere.'

—Sam Knights, writer, actor, activist

'A much-needed book whose time has come. The continued advertising of high-carbon products at a time of climate crisis is a form of insanity. The authors are absolutely right to call for a ban, and it can't happen too soon.'

—Bill McGuire, Professor Emeritus of
Earth Sciences, University College London

Badvertising

Polluting Our Minds and Fuelling Climate Chaos

Andrew Simms and Leo Murray

First published 2023 by Pluto Press
New Wing, Somerset House, Strand, London WC2R 1LA
and Pluto Press, Inc.
1930 Village Center Circle, 3-834, Las Vegas, NV 89134

www.plutobooks.com

British Library Cataloguing in Publication Data
A catalogue record for this book is available from the British Library

ISBN 978 0 7453 4914 5 Hardback
ISBN 978 0 7453 4916 9 PDF
ISBN 978 0 7453 4915 2 EPUB

This book is printed on paper suitable for recycling and made from fully managed and sustained forest sources. Logging, pulping and manufacturing processes are expected to conform to the environmental standards of the country of origin.

Typeset by Stanford DTP Services, Northampton, England

Simultaneously printed in the United Kingdom and United States of America

Contents

List of Figures

Preface

Why did we create a campaign called Badvertising (apart from entertaining ourselves with the name), and why have we written a book about it with the help of some wonderful colleagues?

Let's just say we both have an issue with our culture and economy's tendency to promote its own self-destruction.

It seems it's not enough that we habitually engage in behaviour that reduces our chances of survival and hastens climate and ecological breakdown, through advertising and marketing we actually encourage ourselves and each other to do so. It's like a commercial equivalent of the toxic goading on social media to commit suicide.

Whether it's wall-to-wall promotion of giant, polluting 'sports utility vehicles' (SUVS) which in urban settings rove around like anti-social killing machines, or airlines encouraging us to fly around the world like there's no tomorrow, and in the process guaranteeing that there might well not be.

That's before you even get to the major polluters of the fossil fuel industry who, if you believed their misleading greenwash advertising, you might mistake for renewable energy companies. And this is a primary purpose for the book. One reason that the climate is changing faster than humanity's response to halt the ecological decline is the vortex of misinformation and misdirection from advertising. It drowns out information about the consequences of the overconsumption it encourages, and, like a cuckoo in the nest, it pushes out from our imaginations the adjacent possibilities about different ways to live and organise the economy.

We have both lived, and are products, of this late stage of debt-fuelled, consumer capitalism run riot. What seems normal, immutable and without alternative is actually a tremendously recent phenomenon that has taken root and spread to dominate in a single lifespan. Ours. The issue is personal.

FOR ANDREW …

I grew up addicted to television, and hankering after stuff. My childhood was hypnotised by the promise of electric car racing sets, model jet aircraft, heavily tooled toy soldiers and action men ready for war. The adverts for them would look comical now. But back then, they were utterly compelling. I was behind the wheel, in the cockpit, crouched in the trenches. It was clear, if I owned these things, my life would be, if not complete, as happy and fulfilled as the smiling kids in the ads. To some degree, what toys you had ranked you in school in terms of how worth playing with you were. As a family we were comfortable but didn't have much cash. Come Christmas or birthdays though, I always had something to unwrap. But then the strangest thing always happened, and I never got used to it. Somehow the reality of the car, plane, soldier was different to the advert – crushingly, disappointingly so. The world you saw in the ad didn't come with the piece of plastic in your hands. There was a huge gap between promise and reality. But, instead of that putting you off the idea of toys altogether, it had the opposite effect, you'd come back even hungrier for more, hoping that next time you'd find the thing to fill the void of disappointment. I didn't realise it at the time, but I was the perfect consumer, dissatisfied and so coming back for more, exactly what the system wanted.

Another thing I didn't notice at the time was what else I consumed as I watched the ads and held the toys. That was the assumption, the unquestionable expectation, that it was perfectly normal to be in a world where towns were chock full of cars, skies with planes, and economies were dependent on arms industries.

The realisation that culture of any kind, and especially commercial culture, was not just a simple messenger, a conveyor of neutral information, came early, reading books as a teenager by Susan Sontag and John Berger. The writings of both revealed how the identities we have, the ideas and beliefs that we carry (the things that we believe are 'normal'), and our subsequent behaviours as individuals and societies are created, nurtured, reproduced and maintained in the art, writing, film and advertising that surround us.

The paper I wrote at the end of my first degree in the mid-1980s, during a period of massive deregulation and privatisation, tried to unpick the cultural rationalisation of humanity's reckless, unbounded ecological looting. I'm still unpicking it. It is extraordinary how often nature is invoked to sell things that are killing nature (item number one in evidence, the car).

Something was in the air. A year or so later the Canadian magazine *Adbusters* was launched, with a nice line in critically reworking adverts to tell the truth – 'subvertising' as it became known.

Because there is always a dominant version of events, that one tends to be reproduced most. Anything that challenges the dominant version, the status quo with its particular balance of power and privilege, gets decried as 'Woke', because it is a threat to those who benefit most from 'how things are'.

At school in the 1970s and 1980s vile kinds of racism, sexism and homophobia were present day-to-day in ways which now would not be tolerated. But to call them out then would make you the subject of attack. Slavery abolitionists and suffragettes were both denounced as snowflakes in their own times. The same is true today for the critics of consumer capitalism.

The realisation that beneath the surface of images – whether a sexist airline advert or a painting of Britain's landed gentry in front of their country houses with peasants happily working in fields – there was a complex set of ideas and assumptions was a revelation to a teenage mind. Here were the encoded foundations that allowed sexism and inherited, aristocratic wealth to appear normal and acceptable, when your sense of justice screamed that they were anything but. It was like discovering a hidden civilisation beneath the ground where you had been standing all along.

In my own small way I threw myself into deconstructing the world around me with the critical new lens I'd stumbled on. I lifted it to look through in multiple campaigns over the years, ranging from the Jubilee 2000 campaign for the cancellation of unpayable majority world sovereign debt, to the unethical practices of supermarkets in how they source products around the world and suck the life out of farmers, suppliers and high streets. The lens often turned the picture of the world upside down. Working on low-income countries' financial debts made me think of rich countries' ecological debt. That became a campaign and book of the same name with an entire chapter about the dominance of car culture and how it was marketed. Thinking about how to shrink the ecological debts of the rich meant grappling with uncontained consumerism and the advertising that fuelled it. But how to find a way in?

In a precursor to this book and the campaign it describes, in November 2004 I wrote a cover feature for the *New Statesman* magazine suggesting that polluting SUVs should carry health warnings like packets of cigarettes. The depiction of an SUV with just such warnings all over it earned the outrage of vehicle makers.

Moving from a specific initiative to reframe how one thing is seen to the general, this book might be seen as an attempt to rebrand advertising itself as 'badvertising'. We think there are strong grounds to do so. A commission we ran on the theme of behaviour change to enable rapid transition gave us the opportunity. It led us to create the Badvertising campaign.

Leo and I got to know each other around the time that the climate campaign 10:10 was created. Leo helped set it up and I was on the board. But, seeing things eye-to-eye, and in one instance eye-to-giant-hamster, we had other projects and ideas that Leo mentions below. In a long conversation about our fantasy campaigns that we thought the world needed, we found a special place for taking on the siren song of advertising. This book is about what happened, and what is going to happen next.

FOR LEO …

As a young child, I had three main ideas about potential career paths I might take. I liked the idea of becoming a politician, like my grandfather, Tony Greenwood, who had been a Labour MP, government minister and Peer, as well as a co-founder of CND. On the other hand, I spent large parts of every lesson in school – from maths to English to the humanities – drawing cartoons in the margins of my exercise books, and fancied my chances as an animator creating films and television series. And I was also drawn to a bastard fusion of both of these fields, whereby I felt I could easily improve on the banal carousel of commercial content on our 1980s television screens by becoming an advertising creative – using the visual arts to delight, persuade and change the minds of audiences.

Disaffection with Blair's New Labour government and close encounters with more radical political analysis during my teens put paid to any aspirations of entering formal politics, or of selling my services to the corporate world to flog more stuff to 'consumers', as the public were increasingly being referred to. But I did become an animator, gaining a first class animation degree and quickly finding work in London's West End's production studios.

This was when I discovered that by pursuing a career in animation, I had in fact chosen to work in advertising by mistake. Working as a freelance animator in the West End, I soon learned that nine out of every ten jobs were working on adverts. London's creative industry as a whole and its many televisual production houses are primarily fuelled by advertising money.

If I had felt disquiet over this from the outset, my discomfort grew to become an untenable cognitive dissonance as I followed a personal journey of discovery into the disturbing and very real threat of civilisational collapse as a consequence of oncoming environmental breakdown. I became closely involved in climate activism in my free time, including taking part in civil disobedience at the G8 in Stirling in 2005. But by day I was going to work making art to serve the corporate interests of whoever would pay the studios that hired me. One day I found myself enjoying the craft involved in animating a line drawing of an elephant and a circus entourage, while simultaneously being ashamed that the client for whom the ad was being made, Scottish Power, had the highest coal content in its generation mix of any major UK power supplier.

I realised I could not continue to use my powers for evil. Happily, I was rescued from this dilemma by acceptance on the Royal College of Art's Animation Directing Masters, a prestigious two-year course which would release me from having to take the corporate shilling while I worked out my next move. While studying there I was hired by Franny Armstrong and Lizzie Gillett to work on the climate blockbuster, *The Age of Stupid* (2009), and my graduation film, *Wake Up, Freak Out: Then Get a Grip* (2008), explaining the science and societal implications of climate tipping points, became a cult hit. Together with Franny and Daniel Vockins, I co-founded the 10:10 campaign, taking on a new career path as a professional climate campaigner which I still occupy today (10:10 is now called Possible, and I have held a director post there since 2014).

But a 'career break' I took in 2010 to have kids brought me into serendipitous contact with the esteemed co-author of this book, Andrew Simms. Andrew, then policy director at the New Economics Foundation, had approached me with an idea for a short film based on a metaphor he had taken from one of his lectures about the limits to economic growth, the Impossible Hamster, which remains one of the best things I think I have done. I drew most of it with my animation partner Thomas Bristow while my baby son (now 13) was strapped to my chest. Andrew and I have worked together on and off on a variety of creative environmental projects ever since, culminating in the Badvertising campaign which this book summarises.

At some level, *Badvertising* is my revenge on an amoral fossil economy which has robbed me of my calling. I miss the simplicity of spending my working days producing art; it is difficult to achieve a flow state through reading policy documents and chairing meetings. But the work must be done!

* * *

That is why we've written this book. Looking around we are delighted and energised at the rapidly growing number of people waking up to, and challenging, the advertising industry. For so long it has oddly dodged the limelight as an enabler of destructive, fossil-fuelled overconsumption. That is now changing. It needs to. The culture of consumerism is not only a climate killer, it corrodes our wellbeing and blots out better ways of organising how we share the planet with each other and the rest of nature. Work on cultural discourse can easily slip down a rabbit hole of obtuse theory (we've been down that hole). But this book is a work of practice, a report from the frontline of trying to trigger change. We hope you enjoy it and want to join the party.

Andrew Simms and Leo Murray
May 2023

Introduction: Advertising and the Insidious Rise of Brain Pollution

[S]cary as it may sound, if an ad does not modify the brains of the intended audience, then it has not worked.

—Tim Ambler, Andreas Ioannides and Steven Rose, Brands on the Brain: Neuro-Images of Advertising, *Business Strategy Review*, 2003

The truth is that marketing raises enormous ethical questions every day – at least it does if you're doing it right. If this were not the case, the only possible explanations are either that you believe marketers are too ineffectual to make any difference, or you believe that marketing activities only affect people at the level of conscious argument. Neither of these possibilities appeals to me. I would rather be thought of as evil than useless.

—Rory Sutherland, President of the Institute of Practitioners in Advertising, 2009

Even if 'new' is its favourite word, Advertising is not new.

At its simplest it is a mechanism for selling us products and services, and inevitably some of those products – like tobacco – are bad for us. But it is now clear that advertising is not just trying to sell us products, which may or may not be good for us, it is selling us ideas that will end up making life impossible for billions of people around the world.

That might sound dramatic for something you silence between TV programmes, try to block on your computer screen or flick past in a magazine. But advertising is expert at getting past your defences and under your radar. Increasingly pervasive and targeted, it knows how to find you. Building a picture of your likes and habits from the digital trail of cookie crumbs you leave online, it seeks out your weak points. The worst thing is that it is priming and prompting you even when you are not consciously aware of it, or even when you are and believe you can reject its influence. Advertising has a thousand ways to appear innocuous, or an alternative and amusing form of entertainment, or a carrier of trusted information by using people or organisations you trust or believe in as its messengers.

Altogether, however, it has two major toxic consequences. Advertising activates materialistic values which corrode our sense of wellbeing, it misdirects on the path we all try to tread to find satisfaction in life and is a peddler of false promises. Second, advertising maintains and nurtures a culture of overconsumption that is currently breaking planetary boundaries and wrecking our ecological life support systems. It is so successful at the aggregate level that we end up thinking of ourselves primarily as consumers and customers with the right to purchase limited only by our ability to pay, rather than citizens with a duty of care to ourselves, our neighbours near and far, and the rest of nature.

At its worst, even in the middle of a climate emergency, it is capable of switching our buying habits so that we change for the worse, buying more dangerous and polluting options when safer, less damaging choices are available. Take for example the rise of the SUV, heavily marketed by car makers so that in just around a decade they went from only one in ten of new cars bought to around half.

But consumer culture is so normalised by ubiquitous advertising that it is hard to see how recent and how strange its extreme, contemporary variant has become. Of course traders have always advertised their wares. But the post 'Mad Men' world of advertising and the consumerism it promotes is of a whole other order. Products that are short-lived, with built-in obsolescence designed to be toyed with then thrown away, fast fashion, plastics and endlessly upgraded gadgets. These things didn't just happen, they were the result of some very specific shifts over little more than two generations in recent decades.

A number of factors particular to Western economies and societies in the latter part of the twentieth century created and then fuelled the advertising-driven overconsumption that we live with today.

Several things came together to drive consumer lifestyles and a dramatic rise of per capita consumption, significantly in Europe and North America. After the Second World War there was a large amount of surplus productive capacity in countries like the United States. The 1960s also saw the peak of oil discoveries and surplus production. Here was a product in search of a market. Advances in plastics technology based on oil derivatives hugely boosted the ability to mass manufacture consumer goods that were increasingly cheap and disposable. At the same time there was huge growth in the media and advertising sectors so there was a way of getting knowledge of the products to market, along with visions of happy consumer lifestyles. Conveniently, financial deregulation also led to a huge new market in consumer credit – things like 'hire purchase', or HP, payments. Where once people might save for months to buy a longed-for special item, now they could just go into debt. The scene was set to sell, sell, sell. But as millions have found since to their financial cost and dissatisfaction, getting on this treadmill led not to endlessly raising levels of human wellbeing, but to debt-fuelled overconsumption that now threatens our viability as a species.

The consequence can be seen in how humanity as a whole, though very unequally, now consumes vastly more resources and produces more waste than the ecosystems in our biosphere can replace or safely absorb. Think about it in terms of how far through the year it is possible to get before humanity's collective consumption and waste starts sending us into ecological debt and breaking planetary boundaries.

The last time it was possible to get through the whole year without going into ecological debt – a bit like getting through the month without burning through your month's pay – was in the 1970s, just as consumer culture was gearing up. One of us (Andrew) came up with a measure for this that is now known as Earth Overshoot Day.[1] It gets measured annually by the Global Footprint Network and has been creeping ever earlier in the calendar – since 2001 it's been moving forward at the rate of about three days per year.

In 2022 it fell on 28 July, meaning that, in effect, for the rest of the year irreplaceable ecological resources were being depleted. Ecosystems can only tolerate such overexploitation for a while before they collapse, and it is a deadly game of ecological roulette when that will happen.

The enabler of this great shift of expectations and attitudes has been the advertising and marketing industries. Advertising is selling us an imagined lifestyle – the premise is that we can only feel we are all living our best lives by flying around the world, driving ever bigger SUVS, eating beef from cattle raised on cleared rainforest, and enjoying a vast array of consumer goods like there was no tomorrow. If advertising succeeds in keeping us on our current consumer trajectory, as said, there may not *be* a tomorrow.

The advertising industry may disagree that they are destroying the world – but that would put them on the horns of a dilemma: do they claim that their ads have no impact? Or, as they claimed during the campaign against advertising tobacco, that they just encourage people to switch brands rather than take up smoking in the first place? If the latter, they would then need to explain the huge increase in overall consumption of, for example, polluting, energy-hungry SUVs.

When scientists and policy experts consider potential pathways to even the narrow goal of 'net zero' carbon emissions (there are many questions about methodological flaws and over-optimistic assumptions hidden in the word 'net' of net zero; it means, for example, a huge part of the target is meant to be met by technologies and methods to capture, store and offset carbon emissions that are either unproven or do not yet exist),[2] models converge on an inescapable conclusion: aggregate consumption of a wide range of intrinsically carbon-intensive products and services must fall in absolute terms if the target is to be met. This is why measures such as phase-out dates for internal combustion engine (ICE) vehicles or avoiding the connection of new homes to gas networks feature so centrally in climate policy frameworks. The inexorable arithmetic of emissions reduction dictates that, for instance, 2035 is now the last possible date when new ICE private cars can be sold to UK residents to allow a net zero economy to be achievable by 2050. Yet there is no plan to stop the advertising industry from continuing to push highly polluting new ICE cars to British customers right up until the day their sales are banned.

The challenge is even more fundamental for air travel, a consumption behaviour which is rare in the global context but commonplace amongst the wealthiest residents of the richest nations, and for which no viable technofix substitute is expected to be available on a timeframe that is meaningful for addressing the climate crisis. It is demonstrably impossible to realise the UK's national climate goals without those who are responsible for most of the environmental damage from air travel doing less of it. Hence the UK government's statutory advisers, the Climate Change Committee, have been telling successive governments that 'Deliberate policies to limit demand below its unconstrained level are ... essential if the target is to be met' since 2009.[3] Open calls to reduce consumption in the United States are fewer, but under the Inflation Reduction Act, President Biden set a domestic goal to reduce greenhouse gas pollution by 50 to 52 per cent from 2005 levels by 2030.[4]

In a context in which we have known for years that new policies will be needed to *reduce* demand for things like flights, it is disturbing to reflect that there has been effectively no official policy debate to date around the role of an industry which exists solely to *increase* consumer demand for the products and services of its corporate clients.

* * *

It isn't clear how many adverts we are *aware* of seeing, at any level of consciousness, but what is clear is that they have an effect whether or not we are consciously aware of blanket online, targeted and 'surveillance' advertising.

Exposure to advertising has been increasing steadily since the 1970s, when it was reported that the average person saw between 500 to 1,600 ads per day.[5] A few decades later, advertising had exploded into everyday life. In 2007, the market research firm Yankelovich estimated that the average consumer saw up to 5,000 adverts a day, and – after surveying 4,110 people – half of them said that advertising was 'out of control'.[6]

Fast forward to 2023, and although there are no official figures, the average person is now estimated to encounter between 6,000 to 10,000 ads every single day.[7] How did our exposure to advertising nearly double over less than 15 years? Mainly the huge growth in spending on digital advertising, which was expected to reach nearly $700 billion worldwide in 2023, almost doubling from 2019.[8]

To survive this onslaught, until more ad-free spaces in our lives can be created, we need to grow a critical shield to protect ourselves. Health campaigners have been fighting for decades to strengthen regulations on tobacco, alcohol, gambling and junk food advertising, and feminist and equality movements critique the role of marketing idealised body shapes or skin tones. Experts on wellbeing argue that the constant pressure to purchase and consume 'must-have' brands undermines mental health and promotes attitudes and values that, if anything, make us feel worse – and also make us less likely to behave in a responsible way towards the environment.

Advertising intensifies the commercialisation of our public spaces, and communities living next to large digital screens have found that advertising billboards can create a subtle reduction in our sense of belonging and place.

Car manufacturers and oil companies lobby to weaken climate action, and airlines are failing to take meaningful action to reduce their overall pollution, while aggressively pushing to increase the number of flights.

Some of the companies involved claim to be cleaning up their operations and going green, but their track records on climate change are often those of delay, missed targets, misleading statements and spreading confusion.[9] In reality there is little change and, in 2020, representatives from ExxonMobil, Shell, Chevron, Equinor and BP met the then UK trade minister for a private dinner in Texas where natural gas was championed as a 'vital part of the solution' to tackling climate change.

It is all horribly predictable. But there is hope, and one of the purposes of this book is to bring that hope to life.

Because we can make a difference, as people across the political spectrum have been doing – not just in places like High Street Kensington, London, where a campaign to declutter streets by removing unnecessary signs and other obstacles started as an initiative by one innovative and determined politician, who chaired the relevant committee (and has since begun to spread around the UK with the support of organisations like Living Streets[10]), but also in the very different city of São Paulo.

In 2007, Brazil's biggest city, led by the city's populist, conservative mayor, Gilberto Kassab, introduced the Clean City Law. The result was a near-total ban affecting billboards, digital ad signs and advertising on buses. Despite the threat to income for the city from advertising revenue, the ban was successfully enforced and changed the face of this huge metropolis, reducing the amount of external lighting and forcing the city to acknowledge the reality beneath.[11]

Several other jurisdictions have also felt strongly about the impact of commercial signage on the aesthetic of their environment and some are concerned about the impact of encouraging limitless consumption. We'll visit those in more detail in the final chapter, but it's worth mentioning briefly here that many US states strongly control public advertising and several have banned billboards completely. Vermont and Maine have been billboard free since the 1970s, Hawaii since the 1920s, and Alaska since 1998. In Paris, rules were introduced to reduce advertising on the city's streets by 30 per cent, cap the size of hoardings and ban adverts within 50 metres of school gates.

According to Worldwatch Institute research associate Erik Assadourian, laws like this are important in combating global warming. 'It's not simply greenhouse gases that cause climate change – it's our consumer lifestyle that causes the greenhouse gases that cause climate change', he said. 'Until we end consumerism and the rampant advertising that drives it, we will not solve the climate crisis.'[12]

The Brazilian Clean City campaign was given huge backing by the public, despite a major campaign by the advertising agencies to get them to oppose it. It is also interesting because of two less predictable side-effects. The first was the growth of innovative graffiti art, so much so that Batman Alley, an area covered in street art, is now listed on Tripadvisor as one of the top places in São Paolo to visit. The second was that, without the billboards, some of the symptoms of various social problems – homelessness and gangs, for example – became much more obvious and were then next on the list for reform.

For the first time, many residents were able to see long-standing *favelas*, or slum-like neighbourhoods, that previously had been blocked from view by billboards. São Paulo's City Hall says 474,000 new, affordable homes are still needed today in a city where about half a million families with an income up to about $1,500 per month are homeless. Occupy movements have become commonplace and some have successfully campaigned for access to land for public housing.

It is also interesting as an illustration of the power of big brands and how they try to circumvent such bans, often with the support of business or city officials concerned about revenue loss. In 2017, São Paolo's then mayor, João Doria, tried unsuccessfully to reintroduce advertising on a series of large LED screens around the ring road, in the name of raising funds for the city. Meanwhile, big brands have used graffiti sponsorship and other creative methods to make sure their products are visible to people on the street. Although the ban remains in place, advertising could return through the installation of information screens, interactive bus stop signs and digital clock displays, many of which could accommodate advertising.

Aesthetic problems with modern advertising are very well known, but what really drives these changes is a rising awareness of the impact on people's wellbeing. This starts early and so it is particularly important to consider the impact of advertising on children, who are even less able than adults to filter out its influence.

Street advertising can't be turned off or otherwise removed from a child's environment. Early conditioning towards eating junk food or drink, to being dissatisfied with body image or looks or lifestyle, and to adopting unaffordable consumer lifestyles show that children need protecting from the pressures of consumerism.[13]

As long ago as 1874, the British Parliament passed legislation intended to protect children from the efforts of merchants to induce them to buy products and assume debt. Some places still regulate strongly against advertising to children: Sweden, Norway and Quebec completely bar marketing to children under the age of twelve. But many others rely on industry self-regulation (the USA), albeit often within a wider legal framework (UK).

Today, it is estimated that advertisers spend more than $12 billion per year to reach the youth market, and that children view more than 40,000 commercials each year. This is having a detrimental impact: young children exposed to advertising are also seen to behave less socially, for example, engaging less with other children. According to Susan Linn, director of the Campaign for a Commercial Free Childhood, advertising erodes children's creative play, which she describes as the foundation of learning creativity and constructive problem solving, both of which are essential to a democratic society.[14]

This cascades down throughout our lives as adults, where the values of materialism and consumerism – and the idea that people are consumers first and citizens second – make us less satisfied with our lives, less able to reduce our consumption and less likely to take up pro-environmental behaviours. Commercial advertising tends to push the so-called 'extrinsic' values of materialism, linked to patterns of related overconsumption which display self-reinforcing and negative dynamics.

In numerous, replicated studies, summarised by Professor Tim Kasser, it has been shown that holding more materialistic values is an indicator for having relatively lower levels of wellbeing (see Chapter 3). Just being exposed to images of consumer goods triggers materialistic concerns, which makes us feel worse, and is linked to more anti-social behaviour.[15]

The success of the São Paolo move to clean up their city, known locally as 'Cidade Limpa', was largely down to the huge local support it received, despite fears that the city would lose some of its colour. Some residents worried initially for financial, pragmatic and aesthetic reasons. Some were concerned the city would not only lose revenue from absent ads, but would have to actively spend money taking down the resulting ghost town of empty billboards. Other critics were nervous that without the veneer of advertisements, the urban environment might look worse rather than better, revealed as a gloomy concrete cityscape.

The law was fought at every stage by those with a vested commercial interest in buying and selling prime public ad space. Some in the business community warned that less lighting (by billboards and walls) would mean more dangerous streets, while Clear Channel Outdoor, one of the world's biggest outdoor-advertising companies, sued the city, claiming the ban was unconstitutional.

Despite all this, the advertising billboard market quickly unravelled as advertisers refused to commit to spending that might be banned. In the end, the law went into effect on 1 January 2007, and businesses were given 90 days to comply or pay the price.

Some unexpected consequences of the ban included the city painting over murals by the world famous São Paulo artists Otavio and Gustavo Pandolfo, who were simultaneously showing at London's Tate Gallery – an act that sparked outrage, but also a constructive discussion about the value of street art. Also, the removal of some large signs allowed people to see into poor squatter areas and previously hidden factories for the first time, the poor conditions becoming impossible to ignore once they were no longer masked by huge advertising hoardings.

But more later about the international ripple effect of early pioneers who went out to clean their public spaces of advertising.

* * *

It was because of stories like these that we started the 'Badvertising' campaign to stop adverts fuelling the climate emergency, including ads for cars, airline flights and fossil fuels. And things are already beginning to change faster. At the time of writing, a swathe of cities from the Netherlands to Sweden, the UK and Australia are introducing bans on adverts that promote climate polluting products and lifestyles.

Badvertising, the campaign, was launched in 2020 as a joint venture between the New Weather thinktank, the climate charity Possible and the Adfree Cities network of activists. The launch included the report *Upselling Smoke* – calling for an end to SUV advertising – which led to substantial UK and international press coverage, including a feature on BBC Radio 4's Today programme as well the *Guardian* and *Forbes*.[16] There were informative media interviews between members of the Badvertising team and the Advertising Association as well as motoring pundits.

Then, early in 2021, the campaign pulled together a number of organisations and individuals active in sport and concerned about global heating to discuss how sport, which was once a prime target for tobacco advertising and has become more recently something of a billboard for big polluters like fossil fuel companies, car makers and airlines, could drop 'high-carbon' advertising and sponsorship.

The campaign has built successful connections with international anti-advertising networks like Subvertisers International and, for example, set up a Badvertising group in Sweden.[17] We work with the groups behind Amsterdam's ban on fossil fuels and aviation advertising, with the French network Résistance À l'Aggression Publicitaire,[18] the Dutch anti-advertising group Reclame Fossilvrij[19] and the Belgian environmental NGO IEW[20] – and with the German NGO Deutsche Umwelthilfe who are leading a campaign against SUV advertising.[21]

The problem is that there are so many angles to this particular cause that it can be hard to work out where to start. We hope this book will help a little.

In the next chapter we look at exactly how advertising gets under your skin, often without you even noticing. Then we go into how it makes us consume ever more 'stuff', before looking at the history of measures to control tobacco advertising, the major precedent that demonstrates how advertising can be controlled to protect lives and reduce the consumption of dangerous, damaging products. After that there are three chapters that walk through the issue in the context of sports and the automobile and aviation industries. Next we look at the current, weak regulatory system surrounding the advertising industry – often under-resourced, relatively toothless and too close to the industry it is meant to be regulating. And finally, we begin to imagine what a world with no, or rather much less advertising might look like, how to get there, and where this is already happening and it is already possible to escape advertising's influence.

1
Badvertising, Priming, Propaganda and Surveillance Advertising

Imagine where we can take you.

—Slogan by EasyJet, Imagine advert, 2021

A few hours from home. A million miles from reality.

—Slogan by British Airways, 2019

Take a look at the Imagine advert for EasyJet.[1] It is creative, calming and beautiful. Yet it exists to encourage people to travel in the fastest possible way to fry the planet.

We are so surrounded by advertising that it can become oddly invisible. As a result, advertisers shout louder and louder for our attention, using ever more sophisticated and invasive techniques and digital media.

Take for example the triple-dipped chicken experiment in pre-lockdown 2019 by the young people's campaign group Bite Back 2030.[2]

A group of UK teenagers were asked to come to a London restaurant, and they were invited to order a dish from a long list of options on the menu, but one of the options had already been comprehensively and subliminally advertised to them beforehand, on their journey and through their social media feeds. Every single person taking part ordered the same dish called 'triple dipped chicken', even though there were over 50 to choose from.

Once they had ordered and the food arrived, researchers told them to open an envelope that had been put on their table as soon as they sat down. Like the reveal in a magician's trick, each card inside the envelope read 'triple dipped chicken', so confident in advance had been the experimenters of their ability to control the young people's choices.

This was a controversial form of advertising known as 'priming'. All those taking part had been subconsciously advertised to over a day. Wall posters, print and radio adverts in cabs, and even targeted posts on their social media platforms showing influencers eating the chicken; these had all been automatically absorbed by their brains even when not knowingly aware of the marketing.

Advertising is the cultural water in which we swim. It shapes our choices and wants, our priorities and what we consider to be 'normal' and part of the good life. More worryingly, substantial research – mainly in the USA – shows that we soak up the messages and manipulations of marketing whether we are consciously aware of them or not.

Findings from neuroscience in recent decades reveal that advertising goes as far as lodging itself in the brain, rewiring it by forming physical structures and causing permanent change.

Brands which have been made familiar through advertising have a strong influence on the choices people make. Under MRI scans the logos of recognisable car brands are shown to activate a single region of the brain in the medial prefrontal cortex related to familiar logo recognition.[3] Brands and logos have also been shown to generate strong preferences between virtually identical products, such as in the case of sweet, fizzy drinks. Under brand-anonymysed testing people responded equally, but when cued by the drinks branding, measured responses in the ventromedial prefrontal cortex and expressed preferences were very different. Researchers conclude that 'there are visual images and marketing messages that have insinuated themselves into the nervous systems of humans'.[4]

Researchers have known for decades that '[S]cary as it may sound, if an ad does not modify the brains of the intended audience, then it has not worked.' How we respond to an advert is 'what the ad leaves behind'.[5] Through a combination of experience and exposure advertising, connected to emotional responses, brands and their logos become more 'mentally available'. This happens through the development of new neural pathways which are reinforced by repeated exposure.[6]

This matters enormously because advertising gets to people from a very young age. Dutch research in 2005 showed that children as young as two years old were already 'brand aware' due significantly to exposure to television advertising.[7] They were already capable of recognising a majority of brands when shown the logos of companies including: McDonald's (fast food), Lay's (potato crisps), Wall's (ice cream), M&Ms (sweets), Cheetos (ultra processed snacks), Duplo (toys), Snuggle (fabric softener). Some of these are perhaps more likely than others, and it is a list including a lot of junk food, but it also contained: Mercedes (cars), Nike (sportswear), Heineken (beer), Shell (oil company) and Camel (cigarettes).

Not only is advertising creating a 'choice architecture' for children's lifestyles which can have a lifelong effect on their habits, health and wellbeing, it can have very immediate effects. In another study looking at the effect of exposure to television advertising, pre-school children were found to recognise more unhealthy food brands than healthy ones. Recognition of unhealthy food brands was found to increase significantly between the ages of three and four years old.[8] And in another depressingly sad insight, the pressures of social conformity to have and consume certain recognisable brands have been shown to operate already with children as young as three. Three-year-old's judge each other based upon brand recognition.[9]

Still other research demonstrates how exposure to different brands can influence the behaviour of people, for example making them behave more or less creatively, and more or less honestly.[10] Customisable tools for neural profiling are now available to test the effectiveness of brands and logos' imprinting on consumers' brains.[11]

Having failed to take meaningful action on climate change for 30 years, with the science hardening and observed negative impacts running ahead of predictions, the gap between action that is needed and the speed of our actual change is widening. Why is the climate changing faster than us? One reason is the mixed messages brought to us by advertising that normalise high-carbon products and lifestyles, in contradiction of climate science. Airlines with no credible climate plan encourage dangerous growth in sales using advertising. Over the past ten years, car manufacturers have also shifted away from selling traditional family cars towards promoting ever bigger, more polluting – but much more profitable – 'sports utility vehicles' (SUVs). The share of these energy-hungry giants of the road in new car sales has rocketed and continues to grow. The industry's drive to persuade us to buy these cars has been so effective that it now threatens to trash our climate change targets all by itself. Just by creating new wants and desires in people.

The first economist to talk about advertising actually *creating* wants was John Kenneth Galbraith in his 1958 book *The Affluent Society*. The key questions that need answering over 60 years on are whether we are being 'tripled-dipped-chickened' all the time now – and how much this is contributing to our collective failure to organise a more sustainable future.

* * *

These days, the ad industry's own research suggests that it does far more than help you and me to choose between raspberry jam brand A and brand B.[12] In countries that spend more on advertising as a percentage of their national income, levels of consumption in households are also higher. Surrounded by more advertising, people are more likely to spend and consume than to save.[13]

The ability of advertising to change social norms and behaviour has been well researched in relation to food and drink. The United States saw a massive shift from the consumption of milk to sweet, fizzy drinks in a relatively short period of time.[14] In the space of 50 years from 1945, Americans went from drinking over four times as much milk as carbonated soft drinks, to consuming nearly two and a half times more (heavily advertised) fizzy drinks than milk.[15] These fizzy drinks tended to be loaded with cheap, subsidised corn syrup that also made them addictively sweet. Researchers point the finger at exposure to TV advertising[16] as 'perhaps the single largest factor' in the 'epidemic of obesity among children in the USA'.

The advertisers' toolbox now includes such 'priming' methods as 'stereotype activation', a well-known phenomenon among social psychologists that was first revealed in Gordon Allport's 1954 book, *The Nature of Prejudice*. The New York psychologist John Bargh found in 1996 that, if you subtly prime a group of people of no specific age range with words relating to the stereotype of being elderly, then the behaviour of the group exposed to connotations of ageing and being old then starts to change.[17] In fact, Bargh claimed, although the work has not been widely replicated, they begin to exhibit stereotypical behaviour for old age. They walk more slowly and actually become more forgetful.

Other studies have clearly demonstrated that merely being exposed to images of fancy consumer goods triggers materialistic concerns, which make us feel worse and behave more anti-socially. Children, for example, exposed to advertising were seen to be less likely to interact socially.[18] Studies also show that simply referring to people as 'consumers' rather than, say, 'citizens', triggers more competitive and selfish behaviour.[19] This can have a huge cumulative effect – not just on our own wellbeing – but on others through how we behave towards them.

Psychology professor Dacher Keltner and his academic colleague, Paul Piff, conducted research showing how image-conscious people who drive the highest social status cars also exhibit the most anti-social driving behaviour on the roads.[20] In their experiment, people driving low-status cars routinely stopped to allow people to cross at crossings, but drivers of BMWs, Mercedes and Prius ignored the pedestrians nearly half of the time.

This has implications for the kind of society we want to live in, how we treat each other, whether we are more or less tolerant, and for the embedding of all kinds of prejudice ranging from body type to ethnicity, class, race, gender and sexuality. This is down to the way we are all being manipulated by marketers to aspire to certain lifestyles, whether they are ones in which private cars are privileged and dominant; or where we are invited not to question frequent flying; or lives defined by being in squeaky clean, heteronormative, consumer family units. Advertising primes us to associate certain kinds of feelings and experiences with a product or brand to make it desirable, and in doing so it invites us to embrace materialistic values, priming us with images, colour or language.

'Priming' implies a subconscious reaction to stimuli that influences our conscious decisions. It works by using associations made in our subconscious, and it is almost always unnoticeable to us. We saw how it worked with the 'triple-dipped chicken' experiment above. While barely, or not at all, aware of it at the time, the target audience encountered sounds, words and images via their social media, radio stations and on transport they took that primed them to choose a certain dish. It was almost frighteningly effective. As a result of being primed, when we encounter the product being advertised, we will have triggered thoughts, feelings and behaviours designed to steer us towards making a purchase. Take the fast food outlet McDonald's, which uses the colour red to associate with and prime for excitement. To that is added the constant repetition of the word 'happy', as in 'happy meal', the use of images of happy smiling children, often linked to party-like imagery, children's entertainers (the clown) and 'presents' (toys as free gifts). All these things come into play when a child sees the 'golden arches' logo, makes all the connections that they have been primed to make, and utters the immortal words, 'I want to go to McDonald's'. Remember here the ability of very young children to recognise brands and experience peer pressure to conform to them, and the way in which exposure to television advertising steers them to less healthy food choices.

The same forces are in play when sporting brands such as Nike or Adidas employ elite athletes to promote their goods and tell you to 'Just Do It', or that 'Impossible is Nothing'. But any everyday runner who has been primed to associate elite performance and athletic achievement with pulling on a pair of Nike or Adidas trainers will soon find out at their local parkrun that you can't 'just do it', and that impossible is indeed, very much something. In those gaps between the promise of happy families and Olympic excellence that have primed your purchases, and the reality of junk food's queasy afterburn and brutal confrontations with your own very real sporting limits, lie the dissonance and ceaseless dissatisfaction of the consumer experience. Multiplied a thousand times with the materialistic values and consumer habits it cultivates, this becomes something that eats away at our collective wellbeing. The effects of priming can also spill over into behaviour.

Red Bull is one of a controversial group of energy drinks. One of its distinctive small, 250ml cans contains 80mg of caffeine and between six and seven teaspoons of sugar. It is marketed at active, young people as an aid to energy, wakefulness and the ability to concentrate. To build its brand and associations Red Bull has sponsored a range of 'high energy', extreme sports from Formula 1 motor racing to rock climbing, cliff diving and air racing, to such an extent that it is almost synonymous with risk-taking sporting activity. A range of health researchers, however, argue that regularly consuming energy drinks like Red Bull is itself a risk to your health. 'Is Red Bull bad for you?', asked University Health News ('expert advice from America's leading universities and medical centres'), then answered its own question with, 'Why you should steer clear of energy drinks'.

The ingredients, it said, were reason enough to conclude that 'energy drinks like Red Bull really are not good for your body'.[21] Red Bull is also often used as a mixer with alcohol, hardly surprising given the highly active, young market it is marketed to. But this can have particularly bad consequences. The drink's marketing slogan is 'Red Bull gives you wings', priming for feelings of elation, lightness and invulnerability. But mixing energy drinks like this with alcohol can mask the signs of alcohol impairment leading to people drinking more, and misjudging how drunk they are. Research shows that people who mix energy drinks with alcohol are more likely to 'report unwanted or unprotected sex, driving drunk or riding with a driver who was intoxicated, or sustaining alcohol-related injuries'.[22] A study of the long-term effect of consuming Red Bull concluded that 'athletes and active persons should avoid the long-term consumption of the Red Bull ED and, particularly, its combination with alcohol'.[23]

Already it is possible to see the range of associations, ideas, images and activities with which the potential consumer is being primed. It's a big, party time, energy high. But could it be that the power of priming is such that even without ingesting the product, the Red Bull brand could alter your behaviour and, at least in the digital sphere, your life chances?

In 2011 research conducted in the United States, video gamers were given functionally identical cars within a game to drive, only the cars' branding was different. The players behind the wheel of the Red Bull branded car displayed the brand characteristics of 'speed, power, aggressiveness and risk-taking'.[24] Compared to the other differently branded cars, 'What we see is that people racing the Red Bull car race faster and more aggressively, sometimes recklessly, and they either do very, very well or they push themselves too far and crash', said research author, S. Adam Brasel, assistant professor of marketing at Boston College.[25] Crucially, the subjects in the study were not aware that they were behaving differently. The changes in behaviour resulted from 'non-conscious brand priming'.[26]

Step back for a moment and ponder the cumulative, unconscious effect that advertising and all its devices might be having on us. At a very large scale, could it be that the resulting influence on humanity's collective behaviour is pushing the planet too far, to the point of crashing?

Now let's consider how much further advertisers can go along these lines, with the help of the latest neurological discoveries.

Take colour for example, like the red in Red Bull. Colours provide you with a hidden language, and to some extent a shared one. You can, for example, colour your logo, website background or only specific elements like buttons or content areas. Advertisers are increasingly aware of their target group and the shared language of colours. Although it is far from an exact science, at least in terms of what the industry thinks it is doing, this is a simple guide to the associations and effects that advertisers are aiming for (not necessarily, of course, what the products actually are):

Yellow – happy, optimistic, clear (as in McDonald's, Shell, DHL)
Orange – friendly, confident, warm (as in Amazon, Nickelodeon, Tango, Gulf)

Red – bold, young, exciting, active, passionate (as in Coca-Cola, Red Bull, Virgin, Nintendo, Texaco, ExxonMobil)
Purple – luxury, imagination, depth (as in Cadbury's, Asprey, Hallmark, Wimbledon tennis)
Pink – love, calm, nurturing (as in Barbie, Cosmopolitan)
Blue – trust, strength, control, dependability (as in Barclays, Facebook, Ford, American Express)
Green – growth, health, generosity (as in BP, Starbucks, Landrover, or the Body Shop)
Multicoloured – openness, playfulness, ease (as in Google, Microsoft).[27]

This is a huge simplification. There are countless shades of each colour and the experience of colour is very personal and context specific. However, colours broadly situate brands and products with feelings and associations that are instantly and subliminally read by people.

In the marketing of food the use of colour has become a fairly precise science used for behavioural control. Laboratory experiments show, for example, that 'changing the hue or intensity/saturation of the colour of food and beverage items can exert a sometimes dramatic impact on the expectations, and hence on the subsequent experiences, of consumers'.[28]

We don't know for sure whether, as the marketeers believe, that when the restaurant decor includes orange it stimulates appetite and makes people feel comfortable and want to stay sitting for longer.[29] A fast food restaurant, however, might take a different approach to speed up customer turnover. Certainly, customers can be manipulated, as the shift of girls from blue to pink and of boys vice versa implies. As strange as it may seem to Western audiences today, it used to be blue for a girl and pink for a boy. The cultural shift in associations was led by a number of American ad agencies in the 1950s.[30] There have also been multiple studies on the influence of colour on human emotions, like the famous study in the 1970s that found that jailed prisoners became less aggressive if their cells were painted pink.[31]

The use of colour to manipulate expectations, experience and behaviour also goes beyond tickling our taste buds to how we are encouraged to consume politics. At the UK Conservative Party conference before the 1987 general election, party managers swapped the background from dark blue to lighter blue when they wanted to change the mood from sombre to joyful.

While the jury remains out as to the precise application and effects of colour in advertising, it has become a standard tool of manipulation actively deployed by the industry in such a way as to unconsciously prime and influence the target customer.[32]

You can also use pictures to prime your customers. For example, research has found that being shown images of happier faces can affect our subsequent behaviour. One study looking at the subliminal impact of exposure to facial expressions – happy, neutral or angry – found that people who had seen a happy face, but for a period of time so short that they were not able to consciously notice it, went on to drink more lemon Koolaid than those who had seen either a neutral or angry face.[33]

The human perceptual system is so 'functionally specialized to process faces' in its environment that the slightest of visual cues – such as three dots arranged in a triangle – will be instinctively registered as a face even by tiny babies.[34] Amazon exploits this predisposition by incorporating a curved 'grinning' orange arrow in its logo. The ends of the line narrow while the middle is wide, to emphasise the geometry of a smile, and the arrowhead echoes the cheek-fold at the edge of a broad grin. As the logo's creator Anthony Biles explains, 'it was rooted in the brand proposition rather than the service and ... born of fundamental semiotics. When we're happy ... we smile.'[35] Coca-Cola also grasped this idea in one of their marketing campaigns, 'Open the happy can' in 2009.[36] It showed the top of a can of Coke with a 'smile' in the opening. The idea was that, every time you opened a can, you got a big, wide grin smiling back at you.

But how much is this just the simple manipulation of colours or facial expressions and how much is it something more cynical?

* * *

The term 'subliminal messages' was unknown before 1957, when marketers claimed its potential utility for persuading consumers.[37] This was mainly thanks to the work of one marketeer, James Vicary, who conducted an experiment – or so he claimed – in a cinema in Old Oak, New Jersey, that year. He said he had increased Coke sales by 57.5 per cent, and popcorn by 18.1 per cent, after exposing nearly 20,000 viewers to subliminal projections telling them to 'Eat Popcorn' and 'Drink Coca-Cola'. Each advert was too short – one three thousandths of a second – to be picked up consciously. But equally, Vicary provided no details or explanations for his results, which made it impossible for other academics to reproduce them.

Despite this, the result was a moral panic in the USA, so much so that the CIA published a report, *The Operational Potential of Subliminal Perception* in 1958, which led in turn to subliminal messages being banned in the USA. The report confirmed that:

> Certain individuals can at certain times and under certain circumstances be influenced to act abnormally without awareness of the influence.

Unfortunately, when Vicary was challenged later to replicate the study, he failed to get the same results. And the story gets stranger still. Some decades later, a researcher called Stuart Rogers went to Old Oak to interview people at the cinema where the experiment was supposed to have been conducted, but, when questioned, the manager reportedly told him that no such test had ever been done.[38] Then, in a TV interview with Fred Danzig in 1962 for *Advertising Age*, Vicary admitted that his original study had been 'a gimmick' – and that the amount of data he'd been able to collect was 'too small to be meaningful'.[39]

Vicary died in 1977, and there matters rested – until the Dutch social psychologist Johan Karremans published an article that, extraordinarily, vindicated Vicary's 'gimmick'. He and colleagues ran two experiments involving Lipton's tea, which concluded that:

(a) priming consumers with the name of a thirst-quenching beverage makes it more likely that they will choose that beverage; and
(b) priming also increases their intention to choose the brand, *but only for individuals who are thirsty already.*

In other words, a consumer will only buy a product if they already had some intention of doing so. But then, intentions have to come from somewhere, and advertising over time can both create a market for new products where none existed before (think SUVs), and also increase awareness of products and position them to be part of aspirational lifestyles. This is the rich soil from which intentions emerge.[40]

Consumers spend millions every year on self-help audiobooks containing subliminal messages – products that claim to improve memory, boost self-esteem and facilitate weight loss. Yet most studies have failed to find any evidence that these subliminal messages actually work. On the other hand, in studies of recordings designed to help improve memory or build self-esteem, many who took part with the hope of self-improvement abilities ended up believing that the tapes had been effective.[41] Were those recordings just offering an illusory placebo effect?

People with large budgets who want to influence your choices clearly think that subliminal messaging and priming works. Such as when Ferrari's Formula 1 cars displayed a barcode that was criticised for subliminally flashing the logo of its sponsor company, Marlboro. The barcode was eventually found to be in violation of the ban on tobacco advertising, and Ferrari were compelled to remove the design in 2010.

A belief in the efficacy of priming is why, during the notorious 2000 US presidential campaign, George W. Bush's team sought to use priming to gain an advantage over Al Gore. The Bush campaign was accused of launching a television ad containing a frame with the word 'RATS' in a scaled-up font size. The Federal Communications Commission (FCC) investigated, but they issued no penalties.

It is also why corporations and their marketeers put so much effort and money into product placement in films and television shows, which, since the 1970s, has been increasingly used as a subtle form of subliminal advertising. A famous early example saw the placement of Ray-Ban sunglasses on Tom Cruise in the hugely successful film *Top Gun* lead directly to a revival in the brand's fortunes. We'll see later how, following the ban on tobacco advertising, the number of scenes of people smoking in major Hollywood films spiked upwards dramatically. The advertising trade certainly believes priming can work, and mounting academic evidence agrees: 'Subliminal priming is effective and can influence decision-making' concluded one recent academic review.[42]

The ability to influence actual behaviours 'highlights some unintended consequences of ambient advertising and product placement', said S. Adam Brasel, co-author of the Red Bull experiment mentioned earlier. 'It's an effect that we as advertisers have not been aware of or have been ignoring. All of these brands that surround us are probably having a greater effect on our behaviour than most of us realise.'[43]

The trouble with most research about individual efforts to prime us is that it cannot control for the fact that we live in an advertising-driven consumer world. One that has already long since conditioned and primed us to hear its messages and think of ourselves primarily as consumers, not citizens with a longer list of responsibilities to act on. The environment we inhabit – where children have been primed to recognise the McDonald's arches logo before they can speak – has had us in thrall to corporate influence for generations.

In short, we are born into communities ready-primed. Not, perhaps, for specific products, but for the big ideas of consumer society. No wonder it is so hard to stop burning up the planet. No wonder, as a report on the Red Bull study put it:

> In a world where ambient advertising swaddles buses in wraparound billboards and product placements in TV, movies, Internet, videogames and other media topped $3.6 billion last

year, the Red Bull effect shows advertising and marketing programmes can push beyond simply making a sale. They can have a behavioural influence that consumers don't expect.[44]

* * *

So what is coming next? The combination of priming techniques and the ongoing migration of human experience to the online world ought to make us nervous, partly because those who dominate the online sphere now know a very great deal about us. In 2019, Shoshona Zuboff published a massive tome called *The Age of Surveillance Capitalism*, where she revealed that – for Google at least – they did not at first realise the sheer power this gave them.

She dated the phenomenon back to April 2002 when, after a series of unexpected spikes across the USA in the googling of the same question about an obscure TV character from the 1970s ('what was Carol Brady's maiden name?'), a Google analyst realised that this was a question posed by the host of *Who Wants to Be a Millionaire?* at 48 minutes past each hour as the show aired at the same time in each successive time zone as far as Hawaii.

It wasn't just the accuracy of the data, it was the ability to predict its use that struck Google's executives. They began to see that the data they dismissed as 'behavioural surplus' was not actually surplus at all: it could be the key to their vast future wealth and power.

From their Nest home control system to rifling through our email correspondence, Google is now dedicated to watching our actions and, if possible, the thoughts we express. They do so not to improve their service to customers – we are not customers, because we do not pay them, and are not regarded as such, Zuboff argues – but to make us the subjects and the objects of research using our own data to profit from us. To transform service users into products by leaching 'on every aspect of human experience', she says. And all somehow in the language and rhetoric of liberation and connectedness.

Yet to use that huge new power and influence – when they now keep records of every one of our searches and our conversations and instructions to Alexa – seems to have the potential to be more than a little threatening. Especially when that data is put into the hands of advertisers.

Even so, to build a meaningful connection with customers, you have to be visible in many channels and – even more important – to track users' journeys across multiple platforms and devices, as we'll see later in relation to surveillance advertising.

That is why the prophets of internet futures use the term 'personalisation' – which means that future advertising platforms 'will allow you to map the entire history of consumers' interactions, get data on all the touchpoints and adjust your messaging accordingly ...', according to Michal Schindler of the cloud-based ad-tracker company, Voluum, to 'allow to offer a consistent and unifying marketing experience to consumers'.[45] This level of intelligence on individuals' patterns of behaviour will enable augmented reality versions of adverts, tailored to the needs and personalities of their users. The vision is for continual messaging telling users what they should or shouldn't buy, plus adverts shown – not just on their televisions – but also by their microwave ovens and washing machines, which can in turn be used to make purchases:

> With 5G, they will face adverts in a much higher quality. Think 8K or 360 video ads, 3D models, 100 MP images and so on. You will be able to buy what you see instantly.[46]

The downside of all this breathless excitement is the prospect of the internet monopolies being able to enforce their ways upon us, given that the 'vast majority' of Facebook's $113 billion revenue came from advertising in 2022.[47] The company tracks in detail the activities of its over 2.4 billion users in order to target advertising, but it is not alone. The online leviathan Google does so too, and you are of course followed across their linked platforms like YouTube and Instagram.[48]

With the introduction of products like the Amazon Echo, Google Home and Apple HomePod, devices are now equipped with the new capabilities. They can listen in on and automatically analyse commands, reminders and private conversations.

How will we survive that kind of tidal wave of intensive marketing with our sanity or our bank balances intact? It seems clear that we, as a society and as individuals within it, need to develop a protective layer to make sure we can resist the power of the adverts thrown at us. Having seen the ways in which advertisers can deftly evade our conscious screening of what they do, that means we need regulation, and to expand those spaces in our lives that can be advert free. It suggests especially that public spaces in which we cannot exercise active choice about what we look at should be ad free. Otherwise we will not be able to operate healthily, with reasonable levels of well-being and environmental responsibility in the world.

This aspect of the personalisation agenda depends on a kind of marketing fanaticism about the idea of brands. In our experience, most people have big issues with most – if not all – of the brands they use. For much of the time, at least consciously, they dislike or are sceptical about them, and they are certainly not committed to them at a deep level.[49]

Nor is it likely that having digitally connected, 'smart' kitchen appliances tell us about more ways in which we are supposed to subsume our interests to those of their brand will improve the relationship.

As things stand now, these emerging technologies have yet to fulfil their potential. The comedian Marcus Brigstocke has written about how strange it is to be bombarded with adverts for wheelbarrows just because he bought a new wheelbarrow – how many wheelbarrows does anyone need, after all? As if one purchase had singled him out as some kind of wheelbarrow collector. Meanwhile, both authors of this book find that we are regularly targeted with online adverts for SUVs and private jet flights, because the algorithms know that so many of our online interactions and social media posts reference them. Yet we are of course the anti-customers of such products and services; paying to insert adverts for them into our timelines and browsers is not going to yield clients a return on investment.

But we can hardly rely on online advertisers remaining quite so incompetent, particularly as ever more sophisticated artificial intelligence takes over from human control to coordinate the algorithms which determine the messages we are shown.

* * *

> If you can influence the leaders, either with or without their conscious co-operation, you automatically influence the group which they sway. But men do not need to be actually gathered together in a public meeting or in a street riot, to be subject to the influences of mass psychology. Because man is by nature gregarious he feels himself to be a member of a herd, even when he is alone in his room with the curtains drawn. *His mind retains the patterns which have been stamped on it by the group influences* ...[50] [our emphasis]

This quotation dates back to the founding text for the current wave of modern advertising, which began a century ago. It comes from a 1928 book called *Propaganda* by Edward Bernays, Sigmund Freud's nephew, and known as the 'father of public relations'. It seems to us a perfect description of mass 'priming' of the kind used in the 'triple-dipped chicken' marketing experiment seen earlier.

It is true that Bernays considered advertising a separate function, but closely associated with public relations, or PR. He just saw PR as the task with a broader and higher purpose. Though his views about the human race do not imply that he had a very high opinion of it:

> But when the example of the leader is not at hand and the herd must think for itself, it does so by means of clichés, pat words or images which stand for a whole group of ideas or experiences. Not many years ago, it was only necessary to tag a political candidate with the word *interests* to stampede millions of people into voting against him, because anything associated with '*the interests*' seemed necessarily corrupt. Recently the word *Bolshevik* has performed a similar service for persons who wished to frighten the public away from a line of action. By playing upon an old cliché, or manipulating a new one, the propagandist can sometimes swing a whole mass group's emotions.[51]

After the USA entered the First World War, the high-minded president Woodrow Wilson launched the Committee on Public Information (CPI), which asked Bernays to work for its Bureau of Latin-American Affairs, based in an office in New York. Bernays referred to this work as 'psychological warfare'.[52]

As Freud's nephew, Bernays knew all about psychology and the chasm between what people think they want and what they actually want.

He later described a realisation that his work for the CPI could also be used in peacetime:

> There was one basic lesson I learned in the CPI – that efforts comparable to those applied by the CPI to affect the attitudes of the enemy, of neutrals, and people of this country could be applied with equal facility to peacetime pursuits. In other words, what could be done for a nation at war could be done for organisations and people in a nation at peace.[53]

It was an important discovery. 'This is an age of mass production. In the mass production of materials a broad technique has been developed and applied to their distribution', he wrote in an article called 'Manipulating public opinion'.[54] 'In this age, too, there must be a technique for the mass distribution of ideas.'

Bernays has had a bad press since he died in 1995 at the age of 103. This is partly because of his treatment by the documentary-maker Adam Curtis.[55] But he also received criticism in his lifetime. He was described by one academic as a 'Pluto-gogue' – defined as '"the voice of the wealthy when they can no longer speak for themselves," the successor of the plutocrat of other days …'[56]

Bernays was also horrified to find out that Goebbels was using his 1923 book *Crystallizing Public Opinion* as a basis for his genocidal campaign against Jewish people in Germany.[57]

In fact, he was concerned about democracy, as he said in his 1940 book *Speak out for Democracy*. He believed that if the forces of light shunned his techniques, then they would only be used by the forces of darkness – which would mean the end of democracy. It was a question, as General Booth of the Salvation Army might have said, of why the devil should have all the best tunes.

'Consent is the very essence of the democratic process', Bernays wrote of[58] 'the freedom to persuade and suggest.'[59]

> When I came back to the United States, I decided that if you could use propaganda for war, you could certainly use it for peace. And 'propaganda' got to be a bad word because of the Germans using it, so what I did was to try and find some other words. So we found the words 'counsel on public relations'.[60]

So what is the problem? It is the idea which Bernays was peddling that, to make the world safe for corporations, opinion has to be manipulated on a huge scale. In *Crystallizing Public Opinion*, Bernays explains how governments and advertisers can 'regiment the mind like the military regiments the body'.

This discipline can be imposed because of 'the natural inherent flexibility of individual human nature'. He also instructed that the 'average citizen is the world's most efficient censor':[61]

> His own mind is the greatest barrier between him and the facts. His own 'logic proof compartments', his own absolutism are the obstacles which prevent him from seeing in terms of experience and thought rather than in terms of group reaction.

So when we remember the role Bernays played in persuading women to smoke – by organising prominent women to do so on the New York Easter Parade in 1929, and by calling cigarettes 'freedom sticks' – or by turning the public opinion of John Rockefeller from greed to aged philanthropy, we need to remember also the role he played in capturing the commitment of Americans to a corporate-managed technocratic state, in opposition to the planet. 'Bernays must receive credit, or blame, for an important shift in the methods used by the larger advertising agencies ...', wrote one former editor.[62]

> A few years ago, advertising agencies devoted their attention to straight advertising … Now they have added research workers (which may be a good thing) and great numbers of thinkers, behaviourists, trend-observers, experts with chart and graph, child trainers, students of sleep and what not.

Bernays may have believed his own good intentions, but he was also responsible for the means by which we have become so hoodwinked, about climate change in particular.

SOMEBODY'S WATCHING YOU …

'In today's hyper-connected world, few things are eerier than being spammed with hundreds of furniture ads immediately after having a WhatsApp conversation with a family member about your impending move', wrote Tracy Branfield about surveillance advertising. 'Or talking to your significant other about craving pizza for dinner, only to see Uber Eats banners popping up everywhere you look.'[63]

Surveillance advertising means that most of the apps you use, plus the websites you visit and what you buy online, are tracked and recorded whenever you are online. Advertisers, myriad digital marketing intermediaries and platforms like Google and Facebook (sorry, 'Meta') follow almost everything you do. They can and do build up a detailed picture of your interests, preferences, voting intentions and more, including your age, race and gender, and where you live. Once enough information about you has been gathered to provide a sufficiently detailed picture, this data is then used for surveillance advertising, which involves sending individuals targeted ads based on their behaviour and activities. This is called 'selling down your pixel', and means that advertisers and their intermediaries think that they have your number.

But it could also serve other purposes, and this is where the whole line starts to blur. Because there is so little transparency, so little regulation, and so little enforcement of regulations, no one really knows where or how this private information is being used and abused. Various scandals have given a glimpse of the kind of abuse already proven to happen, such as Facebook's involvement with the controversial consulting firm, Cambridge Analytica, which harvested the personal data of millions of Facebook users, which was then apparently used in the political campaign to elect Donald Trump as president of the USA.[64]

By delivering specific ads to consumers who match certain profiles, surveillance advertising has been accused of depriving people of the power of choice and free will. Also, with huge amounts of personal data being sold to the highest bidder, consumers are increasingly vulnerable to data exposure, identity theft and other similar activities.

And there is another way in which Facebook has become culturally toxic. It is what you might call an emergent property of a system designed to be as attractive as possible to advertisers and therefore sell advertising. Everything on Facebook is curated. To be appealing the platform has to show that people are engaging with it. To maximise the number of people who daily visit the site, Facebook's algorithms promote content that they have learned is most likely to engage. It works: around a quarter of the global population, just over 2 billion people, go on Facebook every day.[65]

Unfortunately, that content tends to be highly emotive, provocative and tends towards the hateful, extreme and often violent. A testament to this is the legion of poorly paid content moderators who suffer post-traumatic stress as a result of their work,[66] efforts which are nevertheless incapable of holding back the tide of content showing humanity at its worst that does get on to the site. The ruthlessness behind the operation of a platform optimised to sell advertising can be seen in the case of a group of Facebook content moderators based in Kenya, who were sacked after trying to organise a union to improve their conditions.[67] Worse still, more positive and peaceful content tends, conversely, to be actively demoted as it is less likely to put advertising in front of users.

Targeted advertising can anyway be harmful, especially for children, who are what Branfield describes as 'pummelled with personalised ads from a young age'. The US online advertising platform OpenX was fined $2 million by the Federal Trade Commission (FTC) for collecting personal information from children under 13 without parental consent. They had also collected information about where the people who opted out of being tracked lived.[68]

One of the earliest – and the widest – examples of tracking user behaviour for advertising purposes came in 1994, when the internet developer Lou Montulli invented a type of file that websites could use to monitor and record information about their users. These became known as cookies.

Montulli's intentions meant well. At the time, the new internet company Netscape was trying to help websites succeed as businesses, and they needed to help them know when people that walked through the door were coming back – and whether they were visiting for the first or the fiftieth time.

Internet users tend to see cookies as a kind of necessary evil – while they are a form of tracking, most people explicitly consent at least to so-called necessary 'functional cookies' – indeed most users are presented with having to accept these as a fait accompli as access to websites is often not granted without first saying yes to them.

On the positive side, they allow people to skip repetitive processes like filling out content preferences or specifying your geographical location. But what most people don't realise is that third parties can use cookies to track their behaviour, not just on a single website, but via multiple websites across the internet.

And since the invention of cookies, companies have found many more ways to record data on their users – and to use the information they gather to gain deeply personal insights into their identities.

The danger of surveillance advertising was set out more than a decade ago by the *New York Times*.[69] They explained how American retail giant Target collected 'vast amounts of data on every person who regularly walks into one of its stores'. It catalogued this information in a database of 'Guest IDs' – unique codes that keep tabs on everything customers buy, their methods of payment, where they live, their ages, names and so on.

'If you use a credit card or a coupon, or fill out a survey, or mail in a refund, or call the customer helpline, or open an e-mail we've sent you or visit our website we'll record it and link it to your Guest ID', Andrew Pole, a statistician who helped design Target's consumer-tracking system, told the paper.

By collecting this information, Target was able to learn about exactly what kinds of products their consumers were likely to buy, as well as when they were most likely to buy them. In fact, pregnant women who shopped at Target were surprised to find that not only did the store seem to know that they were expecting children, it also knew exactly when their babies would be born. Because the store knew what kind of products women buy when they reached their second trimester. This echoes the power that giant UK retailer, Tesco, realised it could wield after developing its own Clubcard points reward system, which built a similarly detailed picture of its customers by tracking their purchases.[70]

Apart from cookies, advertisers use the following, neatly summarised by the Consumer Federation of America:[71]

- *Browser fingerprinting*: This distinguishes a device or browser using things like the device's hardware specifications and its signal. Compared to cookies, fingerprinting is harder to detect, leaving no trace on the device.
- *Mobile tracking*: Extraordinarily, your mobile phone, whether iOS and Android comes preloaded with a unique advertising ID. It's there expressly to enable surveillance advertising. If you know how, you can reset the ID, but advertisers have ways to get around that and still track you using 'probabilistic matching'.
- *Location tracking*: Several apps won't even work without 'location' turned on in your phone, but then many apps that don't even need to know where you are also track GPS location data, which can then be sold on.
- *Probabilistic matching*: Think you've gone under the radar by resetting your device ID? Think again. The sheer volume of data a life online generates, and the fact that companies know most people have multiple devices, means that by comparing data on search histories and other activity across two or

more devices in similar locations means you can be found, they simply work out that the same person is likely to own the different devices.

- *Constantly evolving technology*: Third-party cookies have already been blocked by Safari and Firefox, and Chrome plans to do so too. Yet, not to be beaten, Google has a new tracking method known as FLoC.[72]
- *CCTV spreads its surveillance wings*: Shops can now use cameras to track consumers and the products they may be interested in as they move around stores, that coupled with evolving facial detection software means the very idea of discrete, or secret shopping could soon be a thing of the past.[73]
- *Ears everywhere listening in*: Examples of inadvertent purchases being triggered by eavesdropping smart devices are the stuff of legend, going viral to the amusement of many – and the deep disturbance of those who understand the full implications. Because they're a reminder that homes are now filled with devices, phones and Alexa speakers that are all capable of listening in and recording conversations.

It is very hard for people to avoid all the tracking and profiling that makes surveillance advertising possible. Consumers can clear cookies on their computers, but not all tracking involves cookies. Ad blockers can stop the code that loads ads from running on browsers and can stop some tracking. Fingerprinting is more evasive however, and some tracking code is not connected to ad placement.

Worse, tracking or ad code can sometimes also be doing something necessary for the webpage to work, and many websites have started requiring users to disable ad blockers in order to access content. If advertising, PR and marketing were born out of the civil and military propaganda wings of government, they have now matured into propaganda's corporate twin, surveillance marketing. The emerging ubiquity and comprehensive nature of surveillance marketing has a reach and influence that few have barely grasped.

And if we are unable to help ourselves by resisting being targeted, and stopping our propensity to overconsume – which we take a look at in the next chapter – then what hope does the planet have?

2
How Advertising Increases Consumption

> You can see that in that we live in ghettos with gates and private armies. SUVs are exactly that, they are armoured cars for the battlefield.
>
> —Clotaire Rapaille, the French anthropologist who played a consulting role in the design and marketing of SUVs, arguing that we are going back to medieval times

> Advertisements are a key feature of 21st century capitalist economies, with their nearly inescapable drum beat encouraging consumption. The advertising industry provides support to businesses by developing messages to entice people to spend money on the panoply of goods, services … Changing the practices of the advertising industry holds promise as a way to address climate and ecological degradation.
>
> —Professor Tim Kasser, *Advertising's role in climate and ecological degradation*, 2020

There is a computerised micro world, put together at Ontario University, which is supposed to measure the materialism of students. It is a game where you have to catch fish to survive and make a profit.[1]

The game starts off with 80 fish, and you know in every 'season' that the ones that survive will spawn and double their number – though never above the original 80. If you catch too many fish, you will wipe them out and bring the game to an end.

So when Canadian college students took part in an experiment in 2015 in which they looked at, described and rated several images – some of non-materialistic, non-luxury items, and some of materialistic, luxury items – it meant there was a way of identifying the most 'materialistic' study participants.

After looking at the images, the participants entered the virtual fishing 'micro world', where they managed a fishery along with three other participants – who were actually computer programs. They were told that they would be paid 10 cents for every fish they caught, then they had to manage the fishery.

Sure enough, compared to participants who had been exposed to non-materialistic images, participants who had been exposed to materialistic images depleted the virtual fishery significantly faster, as they showed less restraint in their fishing behaviour in their efforts to make money.[2]

It was one of the more interesting and innovative designs for experiments to test the materialism and acquisitiveness of young people around the world, of several conducted over the past 15 years or so. In 2013, two psychological researchers – Professors Joan Twenge and Tim Kasser, from the University of San Diego and Knox College in Illinois – managed to get hold of measures of materialism among 12th grade graduates of American high schools back to 1976. That meant about 350,000 people at the age of 17 or 18.

How were they defining materialism? 'The importance of money and of owning expensive material items', they said, by way of example.[3]

Having measured the attitudes of American school leavers, they then turned their attention to the total spending on advertising in the USA – and the percentage of US spending on marketing and advertising in a given year when those children were growing up. Sure enough, they found a strange symmetry about the two – a clear link between how much advertising was effective in a given year, and how materialistic teenagers were both that year and five years later at the age of 22 and 23.

This is what they concluded:

> Societal instability and disconnection (e.g., unemployment, divorce) and social modelling (e.g., advertising spending) had both contemporaneous and lagged associations with higher levels of materialism, with advertising most influential during ado-

lescence and instability during childhood. Societal level living standards during childhood predicted materialism 10 years later. When materialistic values increased, work centrality steadily declined, suggesting a growing discrepancy between the desire for material rewards and the willingness to do the work usually required to earn them.[4]

Then there were the Dutch children who took part in a longitudinal study (i.e., research that follows its subjects over a long time) in 2014. At first contact, they reported their exposure to advertisements, and then – one year later – they reported their desire for advertised products and the relative strength of their materialistic values. Their exposure to advertisements had a significant correlation with greater materialism a year later. Mediational analyses – which means that a researcher can test whether the relationship between two variables is statistically explained (or mediated) by a different, third variable – showed that this relationship was explained fully by their desire for advertised products.

In fact, it was becoming increasingly clear that exposure to advertising did lead to more desire for advertised products, which, in turn, led to higher materialism. There was a self-reinforcing dynamic between exposure to advertising, holding materialistic values and the desire to acquire.[5]

By then, young French adults had self-reported their materialism and viewing of television in another study – and TV viewing was significantly correlated with materialism again because of the commercials one can't help seeing.[6] This harked back to an amazing piece of American research by Jeffrey Brand, from Bond University in Australia, and Bradley Greenberg from Michigan State, in 1994.[7]

Brand and Greenberg recruited US high school students from two types of school districts: those with Channel One – a company that broadcasts news and paid for advertisements, into classrooms a few times every week – and those without Channel One. School districts were matched on other important variables, like their socio-economic status.

By now, readers may be able to guess the end of this story: when the students reported their own consumer-oriented attitudes, they found that children in Channel One school districts were significantly more consumer oriented than those in non-Channel One school districts.

Similar stories have emerged from studies in Singapore and China, where undergraduates and adolescents self-reported their materialism and exposure to advertising. Here too, advertising exposure was significantly positively correlated with materialism.[8]

It should be no surprise then that the same turns out to be true of children in the UK. In 2007, Professor Agnes Nairn and colleagues compared UK children, to find out if watching TV was associated with levels of engagement in consumer culture. Again, TV viewing was significantly positively correlated with materialism.[9]

There has been a great deal of research about young people and materialism in recent years and, in summary, three things stand out.

First, most measures show that young people who are more materialistic are not as happy as their non-materialistic friends and contemporaries. This appears to be the case in every culture. Second, the link between materialism and television consumption, which is saturated with advertising, is clear. This was what Juliet Schor, author of *The Overworked American*, revealed back in 1991.[10] People in Australia, Bosnia and Herzegovina, Germany, Egypt, Turkey, Korea and the USA all self-reported their levels of materialism and television viewing, and perception of the presence of materialism in advertisements, revealing a powerful correlation, despite some variability (it was not an absolutely universal finding).

Third, and vital, was the link between materialism – and therefore advertising – and unsustainable consumption behaviour.

* * *

One of the key bones of contention in the long campaign to ban the advertising of tobacco was – as we shall see in the next chapter – whether advertising actually increased consumption, or whether it simply moved the loyalties of existing customers around between different brands of similar products.

When it suits them, advertisers usually claim that brand-switching is all they are in the business of doing. So as part of the Badvertising campaign, we asked Tim Kasser, the professor behind some of the leading materialism research, to see whether and how advertising might cause climate and ecological degradation, directly or indirectly. Professor Kasser found four distinct ways that advertising indirectly causes such harm.[11] He went through the scientific literature about materialistic values and goals, the consumption-driving work and spend cycle, and the consumption of two illustrative products – beef and tobacco – exploring how both are encouraged by advertising and are implicated in causing various forms of environmental damage.

The report he wrote for us – *Advertising's role in climate and ecological degradation* – is a classic of understated logic. 'It seems likely that similar dynamics occur for other products, services, and experiences', he wrote. 'This body of empirical evidence therefore supports the conclusion that if humanity hopes to make progress in addressing and reversing climate and ecological degradation, it would be prudent to rein in and change the practices of the advertising industry.'[12]

There is hard evidence that the actions of the coal, oil, gas, automobile, airline, chemical, plastics and agricultural industries – and others – directly cause climate and ecological damage, sometimes through their greenhouse gas emissions, sometimes through their pollution of air, water and soil, and sometimes through their destruction of forests and other life-sustaining ecosystems, said Kasser.

Then there are the banks that provide the capital that companies need in order to explore for oil, to engage in hydraulic fracking, to build factories that churn out plastics and polluting cars, and so on. There are also mutual funds or unit trusts which offer investors the opportunity to purchase shares of companies that engage in unsustainable actions, enhancing the financial value of those companies. But 'there is another industry that is an indirect source of Earth's current climate and ecological crises but that has, so far, mostly escaped accountability for its actions', said Kasser. 'This is the advertising industry.'[13]

* * *

The most important factors leading people to internalise messages from their environment which suggest that happiness and a good life depend upon wealth and consumption include 'social modelling'. For example, when people see parents, siblings and peers act in materialistic ways, they are themselves likely to imitate those social models.

Media messages are an extra source of materialistic social modelling, because the profitability of both traditional and social media companies normally depends on revenue obtained from presenting users with advertisements that encourage consumption. So it is hardly surprising that many studies find that materialism levels reflect how much people view commercial television and their exposure to advertising.

'Values and goals aimed at maximising one's wealth, status, and image tend to be in conflict with prioritising the transcendent, "larger-than-self" aims involved in caring about the environment', wrote Kasser in 2016.[14] And the human consequences of buying into these materialistic values and goals encouraged by advertising are, as we've seen, lower levels of personal wellbeing, experience of conflicting interpersonal relations, and engaging in fewer pro-social behaviours. Heightened materialisam even undermines their academic or work outcomes.

Most dangerously for our planetary prospects, many empirical studies show that the more that people prioritise materialistic values and goals, the less they espouse positive attitudes about the environment – for example, by saying that they don't care much about environmental degradation – and the less often they engage in pro-environmental behaviours. For example, they tend to be less likely to recycle or to vote for pro-ecological politicians, and they are more likely to live in large houses and to drive gas-guzzling motors.

As we saw at the beginning of this chapter, support for these conclusions comes from a wide range of studies that include children, adolescents, college undergraduates and adults from a variety of nations around the world. They have also been reported when investigators use various methods and means of testing the hypotheses – from longitudinal and experimental studies, through to examinations of regional differences in materialism and energy use. The fact that the negative associations of materialism with pro-environmental attitudes and behaviours is replicated consistently – albeit with some variation – is probably because extrinsic values and goals aimed at maximising our wealth, status and image tend to be in conflict with prioritising the intrinsic, transcendent, 'larger-than-self' aims involved in caring about things that appear beyond ourselves, such as the environment.

Put a little differently, Kasser suggests that 'in order to successfully pursue one's materialistic values and goals, one by necessity has to consume, consume a lot, and consume products, services, and experiences that convey status and an appealing image; such actions are usually difficult to reconcile with living in an ecologically-sustainable fashion'.

Two other studies have looked explicitly at the issue of whether advertising has an indirect relationship with ecological degradation by statistically testing what are known as 'mediational' metrics. In samples of both American and Peruvian adults, results suggested that commercial television viewing is associated with less concern for ecological degradation, and that this association occurs, at least in part, because television viewing leads to high levels of materialism, which, in turn, lead people to care less about the planet.

THE WORK AND SPEND CYCLE

People who live under consumer capitalism are subjected to a whole range of pressures to work long hours. Some of these pressures come from low pay and high housing and energy costs, also from people's employers and peers, some from religion and gender roles. But another source is our conditioned and 'primed' desire to overconsume.

We have had decades building a huge financial infrastructure which encourages people to go into debt in order to pay for the products, services and high-status 'positional goods' that they are encouraged to want. And to pay off those debts, they have to work more hours in order to earn more money that can be spent on consumption. This is the 'work and spend cycle' at the heart of an economy addicted to growth.

Advertisements don't usually promote the work and spend cycle explicitly. Although the pharmaceutical company, Reckitt (now Reckitt Benckiser), behind the cold and flu treatment, Lemsip, did once run an advertising campaign that encouraged people to go to work even if they were ill.

It posed the question, 'What sort of person goes to work with the flu?' and offered the unsettling answer: 'The person after your job.' Then it signed off with the advice: 'Stop snivelling and get back to work.'[15]

Even so, adverts do routinely implicitly encourage the psychological and behavioural attributes that lock us into 'work and spend', and to holding a materialistic world view. Day-to-day, they appear simply to promote specific products (like cars, fast food and sweet fizzy drinks), services (like comparison shopping websites and home delivery) and experiences (like visiting a theme park in another country) that are available in the marketplace. But in stimulating people's desire for and their eventual consumption of 'stuff' and a range of services, exposure to advertising may be another source of long working hours.

Only a few academic studies have so far tested this hypothesis, and those rely mainly on data from the USA and the UK, but what evidence there is supports the conclusion that advertising is playing a role in increasing people's work hours.[16] The fundamental explanation given by researchers who study this topic is that advertising leads people to put a higher value on consumption of what they see advertised and lower value on having more time available for non-work activities, such as time with friends and family, or volunteering, or in the garden, taking a walk, learning a language or doing something creative.

'To put it another way, in the presence of a large volume of advertising, many people come to want to work, shop, and consume relatively more than to rest, recreate, and relate with others', wrote Kasser.[17]

In the early 1980s, two researchers, John Brack and Keith Cowling got hold of data for the years 1919–76 on the average weekly hours of work per production worker in manufacturing in the USA, and the annual real expenditure on advertising there. They also collected data on control variables – wages, for example. Holding control variables constant the annual amount of real expenditure on advertising reflected production workers' average weekly work hours (in researcher speak, the two were 'significantly positively correlated').[18]

The decision to work more hours rather than spend time in other ways comes at a price. For example, compared to those who decide to work shorter hours, those who work longer hours have lower physical and psychological wellbeing, as well as less satisfying interpersonal relationships. Many studies also show that long working hours are associated with higher ecological footprints, greenhouse gas emissions and overall energy consumption.[19]

These findings generally replicate whether analyses are conducted on individual people, on households, on US states or on nations. The only proviso is that some evidence suggests that high working hours are more damaging to the environment in more economically developed nations, like the UK and USA than in less economically developed nations.

Two explanations have been put forward for the positive association between work hours and ecological damage. The *scale effect* suggests that when individuals work more hours, they earn more money that they then spend on consumption; scaled up, many people working many hours leads to high levels of overall economic and ecologically damaging activity.

The *composition effect* suggests that when people work long hours, they have less time to engage in ecologically sustainable activities that are relatively time intensive, like riding a bicycle instead of driving a car or growing food instead of buying it pre-packaged at the grocery store, and preparing and cooking fresh ingredients rather than buying takeaways and ready meals.

Clearly these two explanations are not mutually exclusive, and there is some empirical support for both.

RED MEAT

Gary Brester and Ted Schroeder were both associate professors of agronomy at Kansas State University in 1995, when they managed to get hold of data from the years between 1970 and 1993 on how much beef, on average, people eat in the USA. They also managed to find out how much money was spent across the country on generic and branded beef. They added in data on variables, like prices.[20]

Sure enough, branded (but not generic) beef advertising was associated with increases in beef consumption.

A similar study was carried out in South Korea more recently by J.H. Cho and colleagues at Pusan National University in 2009, using data on monthly beef consumption there from January 2004 to November 2008, and who also obtained information about spending on generic beef advertising expenditures plus other possible variables, like prices and food safety.[21] Here, generic beef advertising too was clearly linked to demand for beef; the more advertising there was, the higher the demand.

Beef has been shown to be uniquely damaging to the environment, relative to comparable foods. For instance, physicist Gidon Eshel, from Bard College in upstate New York, and colleagues used data collected from US federal agencies to estimate the land, irrigation water, greenhouse gas emissions and nitrogen requirements of each of the five main animal-based categories in the US diet: dairy, beef, poultry, pork and eggs. Results show that beef is by far the least efficient of the five animal categories across the environmental metrics examined. Compared to an average of other animal sources of food including dairy, poultry, pork and eggs (which are relatively similar in environmental cost), to produce 1 megacalorie (1 million calories) worth of beef requires, on average, 28 times more land, eleven times more irrigation water, five times more greenhouse gas emissions, and six times more nitrogen.[22]

As such, beef consumption has been shown to be associated with levels of both generic and non-generic advertising of beef products, and to result in relatively extreme climate and ecological degradation. And, of course, consuming beef has been associated with other negative health outcomes ranging from Creutzfeldt Jakob Disease, to heart disease and colorectal cancer.

Many studies make it clear that raising cattle for beef relies on unsustainable water usage, causes destruction of habitat-providing and carbon-capturing forests, and emits high levels of both greenhouse gases and chemicals, such as phosphorus and nitrogen, that cause excessive growth of algae in bodies of water. Beef production, much more than logging for timber, has been identified as the primary cause of tropical deforestation. The evidence is clear that through encouraging the consumption of beef, advertising indirectly causes climate and ecological degradation.

TOBACCO

Tobacco is another product that throws a stark light on many of the societal issues around advertising. The history of the struggle to end the promotion of smoking and tobacco products has some uncanny parallels with the challenge today of stopping adverts that fuel the climate emergency.

We look at this in much more detail below; the next chapter tells the extraordinary story of a campaign over four decades to stop cigarette advertising and marketing in the UK. But here it is good to note how the tobacco industry knew, but suppressed the science on the harm the product caused, lobbied fiercely against the regulation of advertising but, in several important ways, was ultimately defeated.

Even that battle is not entirely won around the world, demonstrating both the cynicism of the industry and the need for constant vigilance, with tobacco companies exploiting looser regulation in parts of the world less wealthy than Europe and North America; with rising numbers of new smokers among adults and children in 'at least' ten African countries, and increased smoking among

13–15-year-olds in 63 out of 135 assessed countries.[23] And, even in the richer parts of the world, battles are still being fought, for example over the introduction of plain packaging, which the industry resists.[24]

Interestingly, there is a body of evidence that the consumption of tobacco has been shown not only to be associated with exposure to advertising, but with climate and ecological degradation.

Not long before lockdown during the Covid-19 pandemic, Su Myat Cho and team from Nagoya University Graduate School of Medicine in Japan, surveyed grade 10 and 11 high school students in Myanmar about their use of any type of tobacco product in the previous 30 days. They were also asked about their exposure to various types of tobacco advertisements – and related questions about things like whether their parents, friends or siblings smoked. Students who reported exposure to tobacco advertising were around six times more likely to report tobacco use in the previous 30 days than were students who reported no exposure to tobacco advertising.[25]

In 2019 a detailed longitudinal study of 11th and 12th grade school students (16–18-year-olds) in the US was published.[26] At first contact, subjects self-reported their tobacco use and the frequency with which they saw advertisements for tobacco products. Two years later, they again reported their tobacco use. Compared to adolescents who continued to avoid using tobacco products, those who began using them in young adulthood were significantly more likely to have reported frequent exposure to tobacco advertising two years earlier. In a much more immediate way, if people who are already smokers view scenes in films with characters smoking they are likely to smoke more.[27] Since tighter controls of explicit tobacco advertising have been in place, the occurrence of smoking scenes in major movie releases has increased dramatically.

Between 2010 to 2018, 'tobacco incidents' appearing in Hollywood's highest earning movies went up by 57 per cent, more sinisterly, that included a 120 per cent increase in films rated PG-13 targeted at young people.[28]

Here it gets even more interesting. The majority of films with these smoking scenes were biographical, so could this be explained by a sudden interest of the movie industry in the lives of people who just happened to smoke, and the needs of historical accuracy? No, because it turns out that about three quarters of the characters smoking in these films were fictional. 'One thing for sure is, these are not random creative decisions', Professor Stanton Glantz, of Smoke Free Media at the University of California, told the *Sydney Morning Herald* for its investigation into the trend, 'Nothing in Hollywood happens by chance.'[29] Until further evidence finds its way into the public domain, however, we will have to make a best guess about what is pushing the trend.

In the USA, tobacco companies have to declare to the Federal Trade Commission that they haven't paid for product placement in movies, TV and also in video games, but streaming content is exempt. And, today, streaming is everywhere.

Given the many physical health risks caused by smoking tobacco, a great deal of research has been conducted on the associations between the advertising of tobacco and its consumption. Although some still claim that the findings are equivocal, the overall body of research linking advertising with the consumption of tobacco is strong enough that Article 13 of the World Health Organization's (WHO) Framework Convention on Tobacco Control begins with the statement that 'a comprehensive ban on advertising, promotion and sponsorship would reduce the consumption of tobacco products'.[30]

As such, the WHO includes enforcing tobacco advertising bans as one of the six components of its core policy proposals – it is the E in its MPOWER programme. And, as Tim Kasser writes:

> Each stage in the life cycle of a cigarette, from growing the tobacco to manufacturing the cigarette to smoking the cigarette to disposing of the cigarette, is associated with specific climate and ecological risks.[31]

OTHER DAMAGING PRODUCTS, SERVICES AND EXPERIENCES

These are just a couple of examples where evidence of the influence of advertising on consumption is documented in the academic literature. But it seems highly likely that similar dynamics occur for other products, services and experiences. For example, it is well established that sports utility vehicles (SUVs), which we also look at in much more detail later, are among the most highly carbon-emitting means of personal transport, and that airline flights, often dubbed 'the fastest way to fry the planet', are a much more carbon intensive means of travel than taking trains or buses, or, indeed, than staying or holidaying at or near your home.

SUVs continue to become an increasingly large percentage of the overall automobile market, and flying has become an increasingly frequent way for people to travel. Two more recent studies also support the idea that the advertising industry contributes to climate and ecological degradation because it encourages the consumption of SUVs and of leisure airline flights.[32]

Researchers recently calculated the carbon cost of an award-winning ad campaign from 2015 to 2017 by the car maker Audi. They combined information on the increase in Audi sales attributed to the new campaign (133,000 cars) with life cycle assessments of the emissions of the cars that were sold. They estimated that the campaign led to over 5 million tonnes of additional carbon being emitted than if the promotion had not happened.[33] It's hardly surprising and merely points out what the industry would want to hear, which is that effective promotions increase sales beyond what would be sold without them.

Another piece of research made the same point about advertising conducted by the aviation industry. In 2020, a German government-funded survey asked around a thousand adults how often they had seen online advertisements and social media posts encouraging consumption of a range of things including clothing and leisure airline flights, about their own aspirations to fly, and how much they had actually bought and flown in the previous year.

What the study found was something that any non-academic would probably have assumed, but good researchers know you have to model. Namely, there was a clear relationship between 'consumption-promoting online content and consumption levels'. Analysis backed the conclusion that exposure to advertising predicted the desire to fly, which, in turn, predicted how much participants reported actually having flown for leisure in the previous year.

Ecologist Vivian Frick and her team of Berlin researchers found that the parallels were so close that it was possible to estimate how much respondents had flown according to how many adverts they had seen.[34]

Much more work remains to be done on the relationships between advertising and the purchase of SUVs and airline flights, to see if findings match the more extensive research on beef and tobacco.

'We cautiously predict that future studies will indeed find similar relationships, and we hope that researchers will take up these questions', writes Kasser.

> In the meantime, given the reasonably large body of scientific evidence ... consistent with the argument that advertising has indirect effects on climate and ecological degradation, and given the scale of climate and ecological degradation that earth's inhabitants currently face, we suggest that the most reasonable conclusion is that advertising also leads to increased consumption of SUVs, airline flights, and other products, services, and experiences that are damaging earth's climate and ecology.[35]

Of course, if some in the advertising industry want to disagree they will find themselves in the curious position of having to argue that their profession is ineffective. Meanwhile, the evidence is stacking up to say, at the very least, that it would be prudent to rein in and change the practices of the advertising industry.

* * *

A strong intuitive case against advertising of high-carbon products and lifestyles exists, but it has evidence on its side too, just as did the campaign for the tobacco advertising ban – a ban now two decades old in the UK. It would be interesting to hear a scientific justification by Shell or BMW or the airlines about how they are not, by promoting oil and energy intensive forms of transport – high-carbon, high-energy products and lifestyles – in fact, undermining society's efforts for a greener, more sustainable world.

These are all reasons to resist the efforts towards making modern marketing even more powerful than it is already. It is also grounds to be suspicious of the few academic programmes specifically focused on *neuromarketing* or *consumer neuroscience*.

In the USA, Temple University, the University of Akron and Iowa State all have active research in the area. Other schools are focused on related fields, like consumer behaviour, neuroeconomics and decision science. Generally, these programmes don't award a degree in 'neuromarketing' per se, but may provide both coursework and research opportunities. Among American academics, neuromarketing has had to overcome the perception that it may be a pseudoscience. Or that, if it's real, it might be evil.

Back in 2007, ScienceCareers.org published an interesting perspective on careers in neuromarketing. The ScienceCareers article played down the immediate career prospects in neuromarketing, noting that much of the action is still in the academic world. They also pointed out the gulf that exists between academia and business – or it did until George Loewenstein and colleagues at the Decision Sciences Department at Carnegie Mellon University carried out a successful study of people's buying habits by scanning brain activity using an MRI scanner.

It appears that a subcortical brain region known as the *nucleus accumbens*, which is associated with the anticipation of pleasure, becomes activated when subjects conclude they have a good deal. Loewenstein's researchers were able to predict whether each subject would buy what they were offered, just by looking at their brain waves. Loewenstein says that he doesn't see any practical applications of his discovery for business – but we are not so sure. It is clear that, on both sides of the Atlantic, car marketing departments and ad agencies have employed psychologists to help them sell cars that would otherwise not be sold. In the UK, the British Psychological Society (BPS) has a clear ethical code, including the following section on 'responsibility':

> 3.3 ... Psychologists value their responsibilities to persons and peoples, to the general public, and to the profession and science of Psychology, including the avoidance of harm and the prevention of misuse or abuse of their contribution to society.[36]

Facilitating the promotion of products that are harmful, and take us away from being able to reverse the climate and ecological emergency would seem clearly to fall foul of that code, and we would urge the BPS to investigate whether their members are helping to promote the products and services of major polluters, in breach of this section of the code, and to act accordingly.

Professions which invite psychologists to work in marketing and, presumably, to manipulate people into consuming more are even promoted within academia. On its website, the famous American Psychology Association (APA) showcases careers in marketing and product development.

That is why we used our 2021 animation, voiced by Dr Chris van Tulleken, to depict the effects of advertising on our mental health, conscious agency and consumption behaviours as a form of 'brain pollution' – intrusive physical changes to the brain which cause the victim to act in ways which undermine their own best interests.[37]

We sought to highlight the conspicuous absence of much-needed public information campaigns on the changes needed to combat the climate crisis, with a little added flavour of vintage, but modernised, public service broadcasting from a fictional government department called the Ministry for the Climate Emergency. The trouble is that actual governments struggle enormously when it comes to doing the right thing. Today we take for granted controls on tobacco advertising and that it's wrong to blow smoke in other people's faces in public places.

But even under a mountain of scientific evidence calling out the public health emergency of smoking, the story behind it shows that action was never a foregone conclusion, and in some places remains an active campaign – as we'll see in the next chapter.

3
How We Banned Tobacco Advertising

> The problem is how do you sell death? How do you sell a poison that kills 350,000 people per year, a 1,000 people a day? You do it with the great open spaces … the mountains, the open places, the lakes coming up to the shore. They do it with healthy young people. They do it with athletes. How could a whiff of a cigarette be of any harm in a situation like that? It couldn't be – there's too much fresh air, too much health – too much absolute exuding of youth and vitality – that's the way they do it.
>
> —Fritz Gahagan, former marketing consultant for big tobacco, World in Action, 1988

> A young adult has the right of free choice whether to smoke or not, but that right to choose does not exist in this country. A young adult cannot pick up a magazine, read a newspaper, watch television or walk in the street, without being bombarded by the tobacco industry. That is not freedom of choice; it's coercion.
>
> —Dr John Dawson, head of the British Medical Association's (BMA) professional division, at their press conference to launch their campaign to stop tobacco advertising, 1984

Jill Craigie was famous for three things – first, her track record as a documentary filmmaker, second, for being the supportive spouse of the Labour politician and biographer Michael Foot, and third, for being a very careful driver. The third of these reputations suffered when, on 21 October 1963, with unexpected positive consequences for the nation's public health, she failed to stop at a red light on a country lane between Abergavenny and Ross-on-Wye.

Also in the car with her was her husband, her grandson Jason, and Vanessa, the family dog. The car was hit by a large Lucozade delivery lorry coming from the other direction. Her hand was crushed, Jason and Vanessa were unhurt, but Michael had all his ribs fractured, and was left with a broken leg and a punctured lung. He wasn't expected to live.

In the event, he left hospital in Hereford a month later, walking with a stick – as he did for the rest of his life – and, miraculously, without his lifelong asthma which had prevented his conscription during the war. Michael had spent his time in convalescence reading Montaigne and E.P. Thompson, and giving up smoking. He had smoked 70 a day for three decades.

The whole experience left him £60,000 richer, thanks to an insurance payment, but also a non-smoker – just as the world was waking up to the consequences of smoking. It also meant that it was to be Foot who, perhaps more than anyone else, demonstrated the moral fix the government was in – that it could no longer reasonably just tax tobacco and turn a blind eye to its effects.

What had happened was that, just over a year before Foot's accident, a group of doctors led by epidemiologist Richard Doll, who was then director of the statistical research unit of the Medical Research Council, published a report via the Royal College of Physicians (RCP), called *Smoking and Health.*[1]

We date the start of the campaign against smoking from this highly influential report. Both the campaigns against smoking and against burning fossil fuels have come to focus increasingly on the role of advertising in promoting products – cigarettes or SUVs – which are deeply destructive of people's lives, not just of those consuming the products but of the people around them. Blanket prohibition of sales of products on which large parts of the population have become utterly dependent, whether physically addictive substances or fossil-fuelled transport, could be seen as unworkable. But an obvious place to begin to avoid and to escape such harmful dependencies through policy is by acting first to prevent them being actively promoted from spreading further through the population.

What was in *Smoking and Health* was powerfully persuasive, but it was hardly new. Doll had known since 1950, when his research into lung cancer patients in London hospitals found that, rather than exposure to motor fumes or tarmac, as he had expected, heavy smoking was the only common factor. In fact, the *British Medical Journal*, which broke the news of these findings, said that of the 1,357 men with lung cancer interviewed, as many as 99.5 per cent of them were smokers.[2]

The parallels between smoking and advertising climate-damaging activities like driving gas-guzzlers are, if not exact, oddly close.[3] Both represent products seeking to create their own marketplace. Tobacco causes damage to the consumers, and tobacco companies benefit from the way that they hook their most loyal customers, with social proof and peer pressure a critical source of new smokers. And while, for example, SUVs are marketed as providing protection for drivers, their physical size, weight and pollution levels create a more dangerous and toxic urban environment for both drivers and pedestrians – helping to compel all road users into an 'arms race' on the streets. Similarities don't end there: where cigarette smoke contains ingredients like benzene, nitrosamines, formaldehyde, hydrogen cyanide, polycyclic hydrocarbons and carbon monoxide, car exhaust has benzene, particulates, nitrogen oxide, polycyclic hydrocarbons and carbon monoxide.

While everyone is affected by vehicle pollution, as with smoking there is also a highly unequal impact in that the pollution and road threat of SUV use is felt first on people who are far poorer. The desire of the vast majority for clean air holds the two campaigns together.

The parallels are strong enough to remind us that – even against the bottomless pockets of the hugely wealthy and powerful tobacco companies – campaigns do manage to achieve their objectives. Starting in Scandinavia in the mid-1970s, and then the UK in 2003, one by one, the places where it was permissible to advertise a product that could kill you began to fall into line. The fact that they did at all was also about the bitter campaigns waged, using innovative tactics, in Canada and Australia.

In fact, the first nail in the coffin for tobacco advertising was hammered in as long ago as 1965, when television advertising of cigarettes specifically was finally banned in the UK (cigars could be advertised right up until 1990). This happened at a moment of opportunity, following the *Smoking and Health* report only three years before – with a new reforming government and a Television Act that had to be passed.

* * *

It was all a long time since Christopher Columbus had been offered 'certain dried leaves' during his first voyage to the Caribbean and Latin America and which, he recorded in his journal, 'gave off a distinct fragrance'. The Spanish conquistadores Rodrigo de Jerez and Luis de Torres are credited with first observing smoking. Jerez became a smoker and took the habit back to Spain.[4]

A century later, virtually the only anti-smoking campaigner was King James I of England. His 1604 diatribe, *Counterblaste to Tobacco*, claimed that smoking was a 'custome lothesome to the eye, hateful to the nose, harmful to the brain, dangerous to the lungs, and in the black and stinking fume thereof, nearest resembling the horrible stygian smoke of the pit that is bottomless'.[5]

The following year, the Royal College of Physicians held a debate on smoking, which dismissed the king's views (doctors at that time were in the habit of prescribing tobacco smoke as a cure for a wide variety of common ailments).

Lung cancer was, however, a rare disease and its increase with the popularity of smoking did not go unnoticed. In fact a link between the two was suggested as early as 1898. Hermann Rottmann, a German medical student, suggested that tobacco dust might be the cause of rising numbers of lung tumours in German tobacco workers. In 1912, another researcher, Isaac Adler, redirected the finger of blame to point at smoking.

And so it continued, until the 1930s, when tobacco brand Chesterfield was allowed to run adverts in the *New York State Journal of Medicine*, describing their product as: 'Just as pure as the water you drink ... and practically untouched by human hands.'[6] Countervailing evidence kept accumulating, but inconveniently for its popularisation, a lot of it was either not published in English or appeared in places that were otherwise discredited. For example,

Angel H. Roffo, director of Argentina's Institute of Experimental Medicine for the Study and Treatment of Cancer, showed in 1931 that tobacco could cause tumours to grow, but published mainly in German and Spanish. Other relevant research suffered because it emerged in Germany when it was under Nazi rule. *Tobacco misuse and lung carcinoma*, by Franz Hermann Muller of the University of Cologne, was one of the first major reports to find a strong link between smoking and lung cancer.[7]

That was 1939. The truth had to wait another decade before British researchers like Doll and Austin Bradford Hill – and American researchers like Dr Ernst Wynder – began to make these connections again. Because of powerful, coordinated and cynical disinformation campaigns by the tobacco industry doubts were perpetuated long after the evidence was convincing, even among doctors many of whom were smokers. So there was never a single 'Eureka' moment of consensus. But already by 1954 the message was somehow getting through, even to the public. A poll in the USA found that 41 per cent of the general public believed that smoking caused lung cancer.[8]

Yet for some reason, in the UK in particular, it was the 1962 report *Smoking and Health* which had really gripped the public. It recommended a complete ban on tobacco advertising, wherever it went, and on smoking in public places.

By the time Michael Foot was giving up smoking, *Smoking and Health* was having very widespread influence. For the first time since the conquistadors, there was a noticeable dip in the sales of cigarettes.

The report was also a big step forward for the Royal College of Physicians, partly because they were such a respectable body. Partly also because most people knew doctors who also smoked, though that was also changing too. Their doctors' study has carried on every ten years ever since, tracking a group of young doctors, which found unequivocal evidence that smokers routinely died before non-smokers, sometimes decades before.[9]

One of those influential people was Charles Fletcher, a professor at Hammersmith Hospital – an elite rower and an unexpected health activist. Fletcher was one of the authors of the 1962 report. He had the full support of then President of the Royal College of Physicians, Lord Rosenheim. People like this thoroughly unnerved the establishment: it was as if their own people were turning on them. As Rosenheim's predecessor, Sir Robert Platt, said during a televised debate: 'What an outcry there would be if a whisky distiller was invited to come on television to say he was not in the least bit disturbed about drunkenness and road accidents.'[10]

What the government did, now under Sir Alec Douglas-Home, was immediately to negotiate a pact between the tobacco and ITV television companies not to advertise cigarettes before the 9 pm watershed – far short of what the 1962 report had recommended.

Swindon Labour MP Francis Noel-Baker opened the battle in Parliament, asking whether – given that the company was spending £5 million a year on television adverts 'persuading people to smoke' – there was really any point in the government's spending what was a tiny fraction of that persuading people to give up? He also questioned if cigarette advertising on TV was even legal under the then current legislation.

Behind both these questions lay a more fundamental issue, and it was Michael Foot who grappled with it during the 1964 budget debate in June of that year. Of course the government should spend more on encouraging people to give up smoking, he said. 'But it is not much good our doing that if, by its deeds, the Treasury shows itself to be quite content with a situation in which our whole financial structure should become more and more dependent upon more and more people's smoking.'[11]

Foot had identified a moral dilemma. It may in fact be there for any 'pigouvian' tax (named after the economist Arthur C. Pigou) – one placed on things that when consumed have harmful consequences on third parties. That is the paradox of governments becoming dependent on tax income from damaging products, even as the tax is used to dissuade consumption. It was all very well for ministers to accept a cosy deal by increasing tax levels on cigarettes but, by doing so, they were impaling themselves on this dilemma

– which put them firmly under control of the tobacco companies. This is what he told the House of Commons:

> Once the Treasury had revealed that it was no longer dependent on such a source of revenue we might get from the government action in other respects such as they ought to have taken already for dealing with this menace. We might then get a government which had the guts to stop all smoking advertisements on television. That would not be interfering with the free choice of anyone. It would interfere with people encouraging others to smoke themselves to death. We might then get a Ministry of Health which, instead of spending a miserable sum – about £25,000 a year on this subject, to counteract the tens of thousands of millions of pounds spent by commercial companies – on persuading people to stop smoking, might do its duty in this respect.[12]

Of course, Foot was also manipulating the parliamentary rules to make his intervention. He was not the health spokesperson, but by framing this as a financial – not to say a moral – issue about money he had been able to demonstrate how governments became complicit in human suffering when all they could do to tackle abuse was to tax it.

So, as they hurtled towards the general election that year, which took them to power, Harold Wilson's opposition began to turn the fears of doctors into a parliamentary campaign. Public health campaigns to dissuade people from smoking were being run in the late 1950s in the UK and gathered pace over subsequent decades. Gradually such campaigns did work. In the USA, where similar health campaigns were run, from that early level of awareness in 1954, by the 1970s, 70 per cent of the public recognised the link between smoking and cancer, rising with neat symmetry to 80 per cent in the 1980s and 90 per cent in the 1990s.[13]

Another champion of tobacco advertising bans emerging in the UK Parliament was Wilson's health spokesman Ken Robinson, Labour MP for St Pancras North. Hansard, the parliamentary record, includes records of his clashes with Postmaster-General Tony Bevins on the subject of television advertising.[14]

Why was the postmaster-general involved? Because, since its inception in 1922, broadcasting was the responsibility of the Post Office. So in 1964, when Wilson unexpectedly took the reins of a minority government later that year, with Robinson as health secretary and Tony Benn (then in an earlier incarnation as Mr Anthony Wedgwood-Benn) as postmaster-general, it so happened that one of the first items on the new government's agenda was a new television act.

The new act made it clear that ITV bosses had a duty to negotiate on advertising with ministers. Some months later, Robinson announced a complete ban on cigarette advertising on television on 8 February 1965 under the previous year's television act. Still under consideration were bans on cigar and pipe tobacco.

There were complaints by the television companies that the government was singling out 'arbitrarily' one form of advertising only. 'If this principle was established', said the Advertising Association, 'no-one could say where it would stop.'[15]

Either way, the first nail in the coffin of tobacco advertising in the UK had been well and truly banged in.

The real difficulty was that the 1962 report had demanded much more urgent measures. And before those could be enacted, it would have to overcome the tobacco industry which was, even then, flexing its considerable muscles in response. They were soon: setting up a fellowship programme for trainee doctors and sending representatives along to cancer conferences; manipulating the idea of media balance to cast enough doubt on the research to keep debate open; and building links with researchers who would be prepared to help them spread the doubt.

Ironically, in the USA, the tobacco companies' own researchers knew as early as the 1960s that the claims about cancer were true, while they were still denying any problem in public. When the US Congress held hearings on the issue in 1965, the industry was ready with their parade of dissenting doctors, many of them funded by their new Council for Tobacco Research. Doctors, dentists and health professionals had for some time been used to endorse cigarette advertising. One now infamous advert for Camel cigarettes features a mother taking her daughter to see a doctor, with the daughter proclaiming 'I'm going to grow a 100 years old!', and beneath her a large headline reads, 'More Doctors smoke Camels than any other cigarette!'

Finally in 2006, a US district judge ruled that tobacco companies had 'devised and executed a scheme to defraud consumers ... about the hazards of cigarettes, hazards that their own internal company documents proved that they had known about since the 1950s'.[16] They were found guilty under the Racketeer Influenced Corrupt Organizations Act. But these tactics had, in the meantime, led directly to four decades of delay, while the industry spread doubt about epidemiological research which showed the dangers of second-hand smoke.

Through the 1990s, the tobacco industry had redoubled efforts to recruit scientists who were prepared to help them spread doubt. They also spread the idea of 'sick building syndrome' to provide themselves with an alternative explanation for health findings, helped along by the Science Environment Policy Project, set up by the PR company ACPO – also planning a second-hand smoke campaign for tobacco giant Philip Morris – to back 'sound science' versus what they called 'junk science' (anything they didn't like).

In the UK, a campaign group FOREST – led by former RAF pilot Sir Christopher Foxley-Norris – was designed to take on the scientific consensus. 'Smoking is only the first ... Beware!' wrote Lord Harris of High Cross.[17] But this was, and remains, a disingenuous form of opposition, because it was not the product itself being banned, but merely its promotion and, increasingly, the ability of smokers to harm others with their habit, through public health measures on where smoking is allowed. Climate campaigners seeking new rules to stop adverts for products and lifestyles that do untold harm face eerily similar obstructions.

* * *

The first result of this new covert, and not so covert, resistance was that, at the end of the year following the ban, ministers asked the tobacco industry to withdraw cigarette coupon schemes. They refused.

The peace movement and flower power was rising and, in 1967, Robinson announced in Parliament the government's intention to 'introduce legislation in due course to take powers to ban cigarette coupon schemes, to control or ban other promotional schemes and to limit other forms of advertising'.[18]

As so often happens, they failed to do this. Instead, they launched the Health Education Council, later to be reorganised as the Health Education Authority, or HEA. Their first anti-smoking campaign was launched: posters asking: 'Why learn about lung cancer the hard way?'

The *Radio Times* – weekly magazine of the BBC – implemented its own ban on cigarette advertising and Finnair claimed to be the world's first airline providing non-smoking accommodation. Otherwise, the chance to legislate seemed to be slipping slowly away. Worse, so many of the abuses remained in place, like Chesterfield-branded chocolate cigarettes – deliberately aimed at implanting the idea of smoking in children. But there was a new generation of physicians emerging who had not discovered with a shock the health effects of smoking after years of being smokers themselves. People like John Moxham in London, or Angela Raffle in Bristol, both now in training and studying the evidence of the health impacts of smoking without having to see through the fog of a lifetime of personal habit.

There was also a great deal of campaigning energy emerging outside the UK, like the BUGA UP campaign in Australia.[19] In Times Square, in New York City, campaigners climbed the lighted advertising hoardings and chained themselves to the Marlboro advert.

The other new factor was the emergence of the campaign group ASH or Action for Smoking and Health. It was set up in 1971, under the wings of the RCP to make non-smoking the norm in society and to inform and educate the public about the death and disease caused by smoking.

ASH was to be a critical factor in the campaign: it had the advantage of being able to say and do things that the royal colleges felt they could not, and yet it was able to coordinate across so many different interests and competing messages. Over a generation, ASH was also able to attract some of the most effective campaigners from across civil society (just as the climate movement has been doing more recently).

One of these was David Simpson, who had been an effective director of Amnesty International. He joined ASH as director in 1979. One of those he met early on and inspired was Dr John Moxham, by then a young registrar at University College Hospital in London.[20]

'I saw ASH as an extraordinary organisation', says Moxham now. 'It was both keen on the science and prepared to get things done. David Simpson was really focused. He wasn't perfect – not everyone got on with him – but he did a very good job. And he got me interested in ASH.' Moxham was also then involved in the British Thoracic Society – a membership-based organisation of health professionals focused on respiratory health and lung disease. He persuaded them to launch a tobacco committee.

But in the mid-1970s, ASH's campaign nightmare was coming true – so-called 'safer' cigarettes. In 1977, the social services secretary, David Ennals, said that his advisory committee had given a limited go-ahead for the marketing of two tobacco substitutes, NSM and Cytrel. These products were incorporated into cigarettes which were still mostly made up of tobacco.

The cigarettes were marketed as 'safer' and both were eventually withdrawn after prolonged and vigorous protests from ASH and other groups on the grounds that the advertising was utterly misleading.

Throughout these years, from the mid-1970s to end of this decade, both sides in an increasingly bitter debate were watching nervously. Norway was the first country to act. A parliamentary committee appointed in 1965 brought together experts from a range of disciplines, and two years later its report recommended a ban on tobacco advertising and health warnings on tobacco products. These became law in 1975. The tobacco side was quick to latch onto figures which seemed to suggest no effect on consumption in Norway. In Finland, they suggested that consumption was coming down before the ban, and that it began to rise again afterwards.

In fact, Finland – once with the highest number of smokers in Europe – had halved the rate of 14-year-old boys smoking within two years of the ban.[21] And over time in Norway, as tobacco restrictions widened, the proportion of people smoking steadily declined.

The importance of numbers was why Simpson and ASH invested such time in their own research – which found that 85 per cent of UK tobacconists were selling cigarettes to children, worth about £60 million a year. At that stage, the industry was spending about £150 million a year sponsoring sports. 'It had to be assumed that the industry's promotional activities were succeeding', wrote the British Medical Association (BMA) later.[22]

It was not until the end of 1980 that the new Thatcher government finally announced its new voluntary agreement with the tobacco industry. Four new health warnings were introduced and more space was allocated to them on posters. The industry agreed to cut its spending on poster advertising by 30 per cent, and also promised to try not to put posters within view of schools, although the clause was vaguely worded.

The sponsorship agreement permitted the industry to raise the prize money offered in sporting events to £6 million. All advertisements for these events would have to carry a health warning. The industry announced its intention to spend £3 million a year on health promotion activities. The agreements were denounced as ineffective by ASH and the BMA.[23]

ASH then vowed that its long-term campaign to see all tobacco advertising banned by law would continue. These were not easy times for tobacco campaigners, yet – behind the scenes – a new energy was emerging that would ratchet up the pressure considerably, by bringing the powerful doctors' union the BMA into the campaign.

* * *

It so happened that, in the summer of 1980 the secretary of the BMA, Dr John Havard, ran into the managing director of British American Tobacco (BAT) on holiday. To his great consternation, the chance meeting was followed by an invitation to visit the BAT laboratories in Southampton.[24]

His first thought was to refuse. How could he be seen hobnobbing with tobacco manufacturers? On the other hand, if he was seen to say no, would he look as though he was too puritanical to accept? He did not want it to be said that they 'didn't even accept an invitation to see what was going on', he said.

The meeting was followed by sending a high-powered delegation, led by the BMA chair Sir Jon Stallworthy, for an equally secret meeting in the BAT boardroom.

References to these two visits were removed from the year's annual report by Havard. But, if the BAT officials felt they had managed to compromise leading doctors, they were wrong. Afterwards, Havard went to see Simpson at the ASH headquarters, meeting each other for the first time. He took with him Pamela Taylor, the BMA lobbyist responsible for recent legislation to make seatbelts compulsory in cars.

Simpson shared recent research findings that a hundred MPs had links with tobacco companies because of factories or other key personnel in their constituencies.

The relationship between ASH and the BMA was nearly torpedoed over a controversial anti-smoking leaflet – and by a row over investments. The BMA's investor partners, Jardine Glanville, had been recommending to doctors that they should invest in unit trusts which happened to include tobacco shares.[25]

When the junior doctors' forum debated the issue in Oxford, the motion to condemn the BMA's leaders was lost by only one vote. Immediately after this second narrow squeak, Taylor suggested a big campaign against big tobacco to Havard. He was keen on the idea.

To repair the bridges with ASH, Pamela Taylor took the BMA's professional division head John Dawson to see Simpson and his deputy Patti White. They all agreed it was an important meeting:

involving the powerful BMA was, said Simpson, 'like the Americans intervening in the Second World War'.[26]

It helped that the government's vast Department of Health and Social Security (DHSS) decided they were going to fund ASH's ongoing anti-smoking work to the tune of a huge £130,000 a year. This was nearly ten times their annual budget at the time, and it meant some years of financial security.

It also meant potentially ending up on the same moral fork as the Treasury did, as Michael Foot pointed out back in 1965, as it accepted increasing amounts of money from the tobacco industry to offset the costs they were causing.

The BMA-ASH meeting hammered out a plan for one of the boldest and most exciting campaigns ever waged. ASH itself agreed to keep a low profile during the BMA's campaigning; in return, the BMA would second a researcher to ASH for two days a week.

And so it was that, in November 1984, behind a big slide of children picking up cigarettes in shop windows, Havard launched the campaign at a press conference at the BMA headquarters. Dawson also spoke to explain the key reason they were acting: that the voluntary agreements between the government and big tobacco were by then considered 'a sick joke'.

The idea was, as they explained, that they were not campaigning against smokers, but against the industry that held them captive. It was an important distinction, but it was more difficult in practice. 'What should have been a simple subject, because we were right, was in fact the most complex subject I have ever encountered', said Taylor later.[27]

The way to maintain the distinction, she felt, was to stay positive. Which is why the October edition of *BMA News Review* explained what they were doing with a big picture of the sky, and the headline 'Breathe!'

Campaigners also collected thousands of blank, black-edged postcards, to be sent by doctors to MPs to tell them every time a constituent died of smoking-related diseases. If they only managed a few of the 270 deaths a day caused by tobacco, they would certainly have an impact.

The first lobby of health ministers was less effective, because the new health secretary was Kenneth Clarke, who was not just a smoker himself, but had a John Player factory in his Rushcliffe constituency.

Margaret Thatcher had, in fact, gone to some trouble to avoid appointing any anti-smoking campaigners or former doctors to the DHSS. In 1982, Sir George Young, parliamentary under-secretary of state for health, was moved to the Department of the Environment after it became clear that he would actively campaign for a ban on tobacco advertising and other legislation to control tobacco use. Patrick Jenkin, the health secretary at the time, was also moved away from the Department of Health.

Even so, the Sims private members bill, based on Laurie Pavitt's attempt in 1979, was published, so progress continued. What nearly undermined the whole campaign was an argument about shares. In those less enlightened days, there was still a sense that where you invest money is an entirely separate issue to your fundamental purpose – the two were supposed to be hermetically sealed – even if one was likely to undermine the other. It echoes current debates about divesting from fossil fuel companies. Because oil companies are such a big feature of the stock market, and because of the way that investment portfolios are built, unless oil and gas companies are actively screened out, many institutions who would publicly call for climate action, such as the Church, get caught out investing in them.

Then, as now, in response to the contradictions, there was pressure to act. Even the BMA was found to be on the wrong side of the line and got rid of its tobacco shares.[28]

Now known as the Red Book, a Health Education Council report on which anti-smoking organisations owned tobacco shares created a furious response, including from potential allies like the Royal College of Surgeons for putting them in the firing line for their investments. 'It's not a perfect world', said their spokesperson.

Two days later, Clarke went on the Channel 4 programme *Face the Press*, and said that banning tobacco advertising would not be effective.

Even so, the doctors' campaign had struck a nerve on both sides. The latest agreement with the tobacco companies was due to expire at the end of 1985, and – before that happened – a high-powered doctors' lobby group, led by the BMA chair Sir Douglas Black, had met the sports minister Neil Macfarlane.

Sport had become the most visible sign that tobacco companies were actively spreading their message linking smoking with health. And watching cricket, for example, meant BBC announcers slavishly reminding viewers that they were seeing the 'Benson & Hedges' cup. It was becoming the most visible symbol of a terrible abuse. The contemporary climate equivalent of sport being used as a billboard for harmful products, we look at in a dedicated chapter on major polluters sponsoring sport (see Chapter 6).

But if the Thatcher government was not moving on the issue of sports sponsorship, they were prepared to move on money for health. Health minister John Patten put aside £500,000 to help women smokers via the Health Education Council, part of a new package of £2.5 million. But it was compared to the £100 million spent on sports sponsorship every year by the tobacco companies.[29] It also fell foul of Michael Foot's moral fork.

Actors were also getting involved in the campaign. *Til Death Do Us Part* star Warren Mitchell withdrew from a production at the Bristol Old Vic in protest against tobacco sponsorship. He was followed by Derek Jacobi, Paul Eddington, Miriam Margolyes and Spike Milligan. Sir Roy Shaw, a former chairman of the Arts Council, backed them in the ensuing argument. 'The idea that the arts will take money from just anyone is nonsense. It would not take money from the IRA or heroin pushers, but tobacco kills more than the two put together.'[30]

The purveyors of tobacco hit back cleverly by trying to change the boundaries of the argument by coming up with a new product called Skoal Bandits. They looked like teabags, but were designed to be chewed.

The government stepped in straightaway to make sure they were not used by children. The BMA said they were staggered they had been allowed in shops at all. Luckily, Skoal Bandits were not destined to catch on.

* * *

In 1988, an American court made history by awarding damages against a tobacco firm to the family of Rose Cipollone, a smoker for 40 years, who died of lung cancer in 1984. Her late husband, and then her family, brought the case against the tobacco industry. The tobacco company launched an appeal against the award and the case carried on for another four years.

The Cipollone product liability case in the USA was finally settled in 1992, after eight years. The decision on the case countered the tobacco industry's claim that such lawsuits could be barred by US federal law, and led to hundreds of subsequent lawsuits. But both industry and health lobbies claimed victory. The tobacco industry because the case demonstrated the difficulty and cost of taking them through the courts, and because the ultimate ruling was not straightforward, carrying multiple caveats. For example, it limited potential future litigants to those who became ill due to smoking prior to the introduction of federally mandated health warnings on tobacco in the late 1960s. But from the perspective of the health lobby things had been fundamentally shifted by the case, because it firmly established in the public debate that the tobacco industry was engaged in a conspiracy of misinformation against the public for the sake of private profit, and at the expense of the premature deaths of thousands. For our contemporary parallel, of particular interest is how the lawyer who led the case against the companies, Marc Edell, focused in particular on the power of the tobacco advertising that he argued was a deliberate and unscrupulous strategy to overshadow the health warnings that by then were required on cigarettes, in order to promote smoking and cultivate nicotine addiction. High-carbon advertising, whether greenwashing oil companies, or glossing SUVs as family-friendly adventure toys, or carefree long haul flights could be seen in much the same way, even with the gradual introduction of, say, compulsory emissions labelling.

Meanwhile, the progress in the UK was continuing. British Airways banned smoking on domestic flights in 1988. The same month, a forensic scientist told the inquest into the deaths of 31 people in the King's Cross Underground station fire the previous year that the fire was probably caused by a smoker's discarded match. That, in turn, led to a smoking ban across London Transport. The Royal Navy announced it was to end its 200-year practice of supplying shore-based staff with cheap cigarettes. The Midland Bank, soon to be HSBC, told the world its premises would be completely smoke free (1992). Virgin Group banned all tobacco advertising and promotion from its companies, at an estimated cost of £2 million over the next five years.

A little later, a poll in March 1989 showed that 79 per cent of smokers thought that National No Smoking Day was a good idea – that had been set in 1984 by ASH on the second Wednesday in March (often Ash Wednesday) and around 5 million smokers would try to give up on the day.[31]

All these announcements allowed the message to drip its way through into people's consciousness. The campaign was still working.

John Moxham, who we met as a young doctor in the mid-1970s, was now two decades older and felt increasingly that 'you have to put your neck on the line'. In March 1986, there he was on the front page of the *South London Press*, in an outpatients' department at King's, looking a little dishevelled, explaining that 415 people in the local Camberwell health district died every year because they smoked. 'You could do that then', he says now. 'There were no PR people in the NHS in those days. I had a district administrator, not a chief executive.'

He set up a direct action organisation called Doctors for Tobacco Law, and took giant cigarettes to the AGMs of tobacco companies. By 1991, he was chair of that group too, selling cigarette packets filled with health information outside the Rothmans AGM in Sheffield.

Yet by the time he joined the board of ASH two years later, the campaign looked as if it was about to suffer a huge reverse.

Kenneth Clarke, now vice chair of BAT, had risen to become chancellor of the exchequer. As if by magic, the Department of Health withdrew their annual grant that was keeping ASH afloat. ASH rapidly ran out of money and, when the chair and chief executive both resigned, Moxham found himself in charge.

The seven staff were put on notice, and the board was reduced to a small rump, including Moxham and publicist Alison Monro. It was a stressful period. 'I remember going to the Department of Health to ask for money and a very superior civil servant telling us we wouldn't get any', he said.

It was the director of the British Heart Foundation (BHF), Leslie Busk – a former brigadier – who saved the situation by providing the money to keep going. They have been ASH supporters ever since. Cancer Research, now a major funder, was also enticed in those days to help. Busk said they would provide the money on the condition that their treasurer should sit on the ASH board to oversee their finances. The BHF treasurer is still there to this day.

These were the dog days of the Major government, and it seemed frustrating that so little seemed to be happening. Health Secretary Virginia Bottomley said firmly that the UK government did 'not believe that stopping tobacco advertising would reduce consumption'. What could be done given that there were so many MPs with links to the tobacco industry, yet at the time, UK opposition was considered to be holding back a European Community-wide ban on tobacco promotion?[32]

What had happened to so unnerve the government that they wanted to bankrupt ASH?

The answer was that the Labour MP and former miners' leader Kevin Barron had put down a private members' bill in Parliament to ban tobacco advertising – and he had not just been given an unopposed first reading, but also an unopposed second reading, having passed it by 227 votes to 17. Campaigners were thrilled. Moxham, in the public gallery, was then very depressed that the government whips made sure it was 'talked out' – the practice, known in the USA as the 'filibuster', of deliberately exhausting the parliamentary time allotted to a bill before it can be taken to a vote. He describes Bottomley now as 'totally sanctimonious'.

It was February 1994, and Kevin Barron was made shadow minister for public health. 'There were over 125,000 premature deaths from smoking in the country at the time. If that was happening because we were at war or we had the wrong rules on the roads, the country would be up in arms', he said. 'There was this idea that starting smoking was an adult decision but most people start when they are very young so it isn't an adult decision.'[33]

The details of the new voluntary agreement on tobacco advertising and promotion, announced in May 1994, were published seven months later. Measures include increasing the size of health warnings on posters and banning tobacco advertising on billboards within 200m of school entrances. At the same time, a parliamentary committee report on sports sponsorship and television coverage recommended that no further sporting events sponsored by tobacco companies should be broadcast once existing contracts have expired. The heritage secretary (yes, responsibility over the years had migrated from the postmaster-general to the secretary of state for heritage), Stephen Dorrell, rejected the idea.

At the same time, the Advertising Standards Authority and other regulators were upholding complaints about a Philip Morris advertising campaign which had claimed that the health risks from passive smoking were no greater than drinking chlorinated water or eating biscuits.[34]

Meanwhile, the tobacco side was beginning to muster a coherent response. The Canadian economist Hugh High, from Capetown University, wrote a short book on smoking for the conservative free market thinktank, the Institute of Economic Affairs (IEA), which had been funded by BAT since 1963. He said that any advertising ban would have to fulfil two conditions: (1) The products must be damaging enough to justify overriding the protection of free speech. (2) Any measures have to be reasonable and justified by the evidence.[35] It is worth noting that these questions did in fact justify the measures that were eventually taken against promoting smoking, just as they also justify banning advertising of climate changing products.

Even so, as High pointed out, bans are not risk free. The effects of the prohibition of alcohol are, he said, 'well-known to even the most ignorant parliamentarian'. The pro-tobacco case argued that advertising cigarettes was about brand-building and encouraging people to switch brands, rather than expanding the market. It would have to be if they were going to win, given the consensus that smoking killed people. But, in fact, even where tobacco has been effectively a monopoly – places like Japan, Austria or Cameroon, where brand competition was non-existent – tobacco advertising nevertheless continued, fatally undermining their own argument. The long-term consequences of resisting the industry led finally, in 2022, to global smoking rates falling for the first time, according to research by the University of Illinois at Chicago. But it also showed how the industry's relentless pursuit of profit means it still seeks out new, growing young markets, and often where regulation is likely to be weak. The Western Pacific and Eastern Mediterranean are seen as growth markets, with rising rates of child smokers in a number of African countries being a major concern. One of the authors of the research, Jeffrey Drope, public health professor at the University of Illinois, commented, 'The industry is still preying on emerging economies in ways that will lock in harms for a generation or more.'[36]

The brand-switching arguments have disappeared temporarily from public debate around advertising of harmful products, but they will probably emerge again in different forms over climate change. The US Surgeon General's report of 1989 said that there was 'no scientifically rigorous study available to the public that provides a definitive answer to the basic question of whether advertising increases the level of tobacco consumption'.[37] Yet research dating at least back to the 1970s had identified an increase in tobacco sales in response to advertising and, as described above, the evidence is there and there was more to what was being said.

Commentators like High leapt on the apparent assertion in the US Surgeon General's report that there was insufficient evidence that advertising had a statistically significant effect on tobacco consumption, using this quote selectively to bolster his own mendacious argument for the IEA.[38] However, the full paragraph in the Surgeon General's report continues: 'The most comprehensive review of both the direct and indirect mechanisms concluded that the collective empirical, experiential, and logical evidence makes it more likely than not that advertising and promotional activities do stimulate cigarette consumption.'[39]

This is of course the opposite conclusion to the one High sought to project for the IEA. In 1992, Clive Smee, chief economic adviser to the Department of Health, published his own comprehensive study of the link between advertising and tobacco consumption. He reviewed 19 studies, mainly from the UK and the USA, correlating advertising spending and total tobacco consumption and he concluded: 'The balance of evidence thus supports the conclusion that advertising does have a positive effect on consumption.'[40] Smee also reviewed the impact of advertising bans that had been introduced at the time. The most significant of those were Norway and Finland where bans had been in place for over a decade at the time of the report. 'In each case, the banning of advertising was followed by a fall in smoking on a scale which cannot reasonably be attributed to other factors', he wrote.[41]

The danger was that researchers were vulnerable both to any admission of balance or uncertainty – and to deliberate overstating of the case. Either would be leapt upon by the other side, where their main purpose was the spreading of doubt. This dilemma was also to become familiar to climate change researchers in the 1990s and 2000s. Of course, even without this confirmation, it would have been hard to prove beyond doubt – especially when the tobacco industry put so much energy into spreading this – the extent to which sports sponsorship by cigarette companies effectively started young people smoking. Following High's pamphlet in 1998, the *Financial Times* claimed that an advertising ban would actually increase smoking.[42]

The 1990s saw the endless repetition of these arguments, as the focus of the campaign on both sides began to shift to the courts, and especially those of Europe.

* * *

By May 1997, there was a new government in the UK under Tony Blair. To the delight of the tobacco campaigners, the new government announced its commitment to ban tobacco advertising and tackle smoking among the young.

Tessa Jowell was appointed as minister with responsibility for public health – the first time that public health has been recognised at ministerial level. The health secretary, Frank Dobson, announced that the government would also be banning tobacco sponsorship of sport, but that sporting bodies would be given time to find alternative sponsors.

It was an exciting moment, but – as climate change campaigners know very well – often it is in these final moments of success when things can go terribly wrong. And something increasingly seemed amiss. For one thing, where was Kevin Barron in the government's plans? The former shadow minister for public health in opposition had been overlooked for a ministerial post – something he believes was down to his campaigning for a ban on cigarette advertising.[43]

What was going on became clear some months into the new government. It emerged that Formula One chief Bernie Ecclestone had donated £1 million to the Labour Party, and it was announced in November that same year that the sport would be exempt from a planned tobacco sponsorship ban. The resulting scandal saw the donated money returned to Ecclestone.

The tobacco companies were also mounting a legal challenge in the high court, on the grounds that the new UK law was based on European legislation then under consideration by the European Court of Justice.

The problem was that the occupants of Downing Street had become nervous of the effects of their sports sponsorship ban. They were aware of just how much store that Formula One was putting in the advertising of tobacco. At the end of 1997, BAT confirmed that it had bought the Tyrell motor racing team – former sponsors: Elf – which would be known as British American Racing. BAT was planning to spend up to £300 million over five years on the new team.

But BAT was also in trouble on the other side of the Atlantic. Back in 1994 the chief executives of America's biggest tobacco companies swore in front of a US congressional subcommittee on health and the environment, known as the Waxman committee after its chair, the Californian congressional representative, Henry A. Waxman, that the nicotine they put in cigarettes was non-addictive. It was their way of again saying that they could not be held responsible for the cancer deaths of thousands of addicted smokers. The moment was filmed and has since become iconic as, one by one, seven white men in suits repeat under oath the lie (their own research telling them so) of nicotine being non-addictive.[44]

Jump forward to 1997, and one month after BAT's Formula One announcement, tobacco executives this time admitted at another US congressional hearing that nicotine was, in fact, addictive and that smoking was harmful. Internal tobacco industry documents released to another American court showed that BAT had known at least 20 years before that nicotine was addictive.

It has since emerged that the same pattern of systematic corporate denial and obfuscation has also been employed by major fossil fuel companies such as ExxonMobil and Shell to undermine the scientific consensus that use of their products is the principal cause of global warming – something their own scientists had confirmed, but kept secret, for decades.[45]

Partly as a result of all this controversy, Formula One's ruling body, the Féderation Internationale de l'Automobile (FIA) announced in March 1998 that it would consider bringing forward the end to all tobacco sponsorship if presented with evidence that tobacco sponsorship encourages children to take up smoking.

This remained the elusive problem for the tobacco campaigners. All the way through the battle, they needed the completely unanswerable evidence of cause and effect. They could say, of course, that children were still getting hold of cigarettes and beginning a lifetime of tobacco addiction that would eventually kill them. By the end of the century, the government seemed entirely to have lost their nerve, and had agreed to let the tobacco industry have at least three years to pause their shift out of sponsorship and promotions.

By then, the Blair government's first term only had a year to run and no prime minister was going to introduce a ban which might be so controversial so near a general election – except in such a way that they could be seen to worry about it but actually to change nothing. And so it was that the legislation was debated and passed the House of Lords, in time to fall when Parliament was dissolved in time for the 2001 general election.

In fact, the 2001 election came and went without any sign that tobacco advertising would ever face a ban.

Salvation for the campaigners came in the shape of Celia Thomas, now Baroness Thomas of Winchester, by then an effective political mind in charge of the Lib Dem whips office in the Lords. After the general election was over, she took Lord Clement-Jones aside and suggested that he introduce the same legislation himself. It would embarrass the government and it might even get passed.

Tim Clement-Jones was a long-standing Liberal and was then the Lib Dem health spokesperson in the Lords. By some coincidence, as well as being a political adviser and lobbyist as his day job, he was also a consultant for the Advertising Association. Once it was clear that he had got into the ballot for private members bills in a high enough position, and that he was preparing to press ahead with passing legislation to ban tobacco advertising, he rapidly lost that contract.[46]

He rose immediately before midday on a wintry Thursday evening, and again to sum up when it was dark nearly four hours later. This is what he said about his critics:

> As a long time civil libertarian, I have examined my conscience over the bill. If I were not so mild-mannered I might resent the raising of this issue. No advertiser has unfettered freedom of speech. The existence of the British codes of advertising and sales promotion, administered by the ASA, recognise that. Furthermore, it is clear that Article 10 of the convention permits restrictions and limitations on the right of freedom of expression which pursue a legitimate aim and are proportionate. The protection of health is one such aim.[47]

The bill passed the Lords and the government then promised to adopt the legislation in the Commons. The new advertising and sponsorship ban came into effect at midnight on 13 February 2003. Gangs of workmen were on hand in the big cities to tear down the billboards.

Most Western and European countries have followed suit since then. Ireland, Norway and then New Zealand in that order banned smoking in public places in 2004. At the same time Philip Morris International offered the European Union $1 billion to avoid lawsuits over claims of illegally smuggled cigarettes into Europe, and in the USA a federal case against the industry over 'civil racketeering' was launched seeking $280 billion of 'ill gotten gains', from money made by selling to smokers who became addicted before 21 years old. The following year in 2005, the Framework Convention on Tobacco Control, a global health treaty, entered into force, ratified initially by 40 countries, through which countries committed, among other things, to ban tobacco advertising. In 2008, perhaps an even greater landmark was reached when France, nation of the languid Gauloise smoker, banned smoking in public places, and Germany, another laggard, started to follow suit with eleven of 16 states doing the same, followed by Turkey. Moving beyond advertising to tobacco packaging, in 2011 Australia became the first country to announce a law imposing plain packaging on cigarettes.[48] Brand logos would be banned and all cigarette packets made a dark green colour. Although once again fought by the industry, the measure became law in 2012, and already by 2015 smoking rates in Australia showed a marked decline.[49] In 2014 smoking rates in the USA also hit a record low of 16 per cent.

But, as we've seen, from a global perspective, the battle remains to be won. The total number of smokers may be down, but big tobacco companies remain hugely profitable – and have managed to use loopholes in existing legal frameworks around the world. By the time the UK sponsorship ban was in place, the Benson & Hedges bistro had already opened in Kuala Lumpur.

But why did it take so long to achieve a ban that was so obviously in people's interests? One way of looking at it is that the campaign was so slow because of a lack of conclusive evidence that advertising actually increased smoking, rather than just – as the tobacco companies claimed – increasing brand share. For the purposes of getting a ban agreed, Health Secretary Hazel Blears used to claim that 3,000 lives a year would be saved – but the figure was disputed and it only had authority because it came from the government.

That may be the case but, equally, the biggest step forward was arguably made within three years of the *Smoking and Health* report which started the whole campaign – so it may be that the main lesson is a political one. The TV advertising ban went through, flawed though it was, within the first few months of a reforming government, before the exhaustion and inertia that tends to set in on new governments of all kinds.

Even in 1997, this might have been possible – but having failed to get the ban through during Tony Blair's first term – it may well not have been passed at all if it had not been for the intervention by Lord Clement-Jones and the Liberal Democrat peers.

Like the tobacco industry, the oil companies on which the car and airline industries depend also knew more about the damage their products caused than they publicly admitted. This is revealed in a report, 'Review of Environmental Protection Activities for 1978–1979', produced by Imperial Oil, Exxon's Canadian subsidiary in 1980, more than a decade before the signing of the UN Framework Convention on Climate Change.[50] This spelt out the level of understanding and awareness in the industry:

> It is assumed that the major contributors of CO_2 are the burning of fossil fuels ... There is no doubt that increases in fossil fuel usage and decreases of forest cover are aggravating the potential problem of increased CO_2 in the atmosphere.

One reason for subsequent inaction might have been the added observation that:

Technology exists to remove CO_2 from stack gases but removal of only 50 percent of the CO_2 would double the cost of power generation.

Like the tobacco industry, oil companies acted to confuse the science, lobbied to prevent regulation and sometimes denied outright that there was a problem. Take this example from Imperial Oil chairman and CEO Robert Peterson who wrote in 'A Cleaner Canada' in 1998 that 'Carbon dioxide is not a pollutant but an essential ingredient of life on this planet.'

Of course, greenhouse gas emissions and other pollutants from oil don't just magically enter the atmosphere, they do so when the oil and its derivatives like diesel, kerosene, petrol and gasoline are burned as fuel – and transport is one of the biggest sources of that pollution.

Transport is now the UK's largest sectoral source of carbon emissions, with road transport alone accounting for around a quarter of our total carbon dioxide output.[51] But sometimes it's hard to tell because the car industry, like the tobacco and oil industries, has concealed the impacts of their products – sometimes by providing misleading figures for the fuel efficiency of their vehicles – and sometimes by outright cheating.

In 2017, it was reported in research commissioned by the group Transport and Environment that the actual fuel efficiency of an average new car when driven on the road was 42 per cent worse than its advertised efficiency, burning far more fuel. One of the world's leading car makers – Volkswagen – was first accused, and then admitted to, illegally fitting devices to a huge number of its vehicles designed to 'cheat' emissions tests. An astonishing 11 million cars were fitted with these 'cheat' devices.

One parallel between tobacco and climate-related 'badvertising' initiatives is that, in both cases, people can feel like the victim of circumstances – they feel unsafe in a smaller car, or they are subject to peer or family pressure to jet around the world, or to smoke. They don't feel it is entirely their fault or their choice, and that is because everything from the social norms that advertising has helped create, to the economic conditions and, for example, transport infrastructure, over which people have little control, make change hard. In other words, to be effective, much more effort needs to go on changing systems to make better choices the default ones, than on merely exhorting individuals to change.

It is also the reason why removing the pressures of commercial advertising to buy products that harm the user and those around them is a common sense policy instrument when it comes to tackling the climate crisis. We do not have 40 years left in which to act to prevent climate breakdown. But the tobacco story reveals that change is possible, and in the final chapter we sum up some of the lessons it gives us.

4
Sports Advertising and Sponsorship: The Great Pollution Own Goal

> It would be very difficult for our fellow citizens to identify with companies whose activity would have a big impact on the environment, based in particular on the massive use of carbon-based energies.
>
> —Paris mayor Anne Hidalgo, denying the fossil fuel giant Total access to sponsor the 2024 Olympics, 2019

Why do advertisers cling to sporting heroes? At the 2021 Australian Open tennis championships, the prominent courtside sponsors included a fossil fuel company, an airline and a car maker. High-carbon sponsorship of sport has, in many ways, replaced once common and now disgraced deals with tobacco companies.

Sport used to rely heavily on tobacco sponsorship until the importance of public health overcame vested interests and largely ended the practice. In 1990, more than 20 different televised sports were sponsored by cigarette brands in the United States alone, and a single tobacco company, RJ Reynolds, admitted in 1994 to sponsoring 2,736 separate sporting events in a year.[1]

Today, the world faces a climate emergency and sport is floating on a sea of high-carbon sponsorship. First, from the melting of winter sports to the flooding of football grounds and the cancellation of flagship sporting events due to heatwaves and air pollution, global heating and the emissions that cause it are a huge problem.

Second, sport itself is contributing to the problem directly through all the emissions linked to it.

Third, direct association with promoting high-carbon products and lifestyles not only contradicts the pledges of climate action that many clubs and sports bodies are beginning to endorse, but it poses an increasing reputation risk to sport, which is meant to represent a better, healthier way of life.

How big is the problem? Tobacco sponsorship of English cricket is now prohibited, but the global sports industry was worth an estimated $471 billion in 2018.[2] Corporate sponsorship in sport is a multi-million-dollar business. Some aspects of it go back a long way. In the early days of pre-Olympic, modern athletics, professionalism was common and many races were sponsored by local pubs to attract drinkers. But, after wrapping itself in an elite, 'amateur' flag, it was only in 1984 that the Los Angeles Olympic Games became the first Olympics to sign a corporate sponsorship deal.

Since then, the sector has gone through significant changes from the time when only a handful of brands were able to use sports for self-promotion. Now sports sponsorship, with its celebrity athletes, huge audiences and associations with vigorous, healthy living, is arguably one of the most important weapons in the advertising armoury. In one quick, non-comprehensive survey, we found a total of 258 sports sponsorship deals in various countries with companies promoting high-carbon products, services and lifestyles.[3]

These deals ranged across 12 different sports categories: football, American football, cricket, the Olympics, tennis, sailing, cycling, athletics, basketball, rugby, golf and motorsport. The deals covered clubs, teams, associations, leagues, federations, races, championships, tours, tournaments and stadiums. Football, understandably as the world's leading global sport, was the most targeted by advertising – with 57 high-carbon sponsorship deals.

Within that, the auto industry is the biggest sponsor. That survey threw up 199 deals. Airlines were second with 63 sports partnerships. The high-profile fossil fuel companies Gazprom and Ineos were also prominent. The car maker Toyota was the largest sponsor with 31 deals identified – followed closely by the airline Emirates with 29 partnerships. One estimate put the value of sports sponsorship globally in 2019 at $46.1 billion.

Middle Eastern airlines especially have positioned themselves as leaders in the global market. Among them, the United Arab Emirates (UAE)-based airline Emirates is top of the list, having signed countless partnerships with football, tennis, rugby, sailing, horse racing and golf clubs around the world.[4]

Around two thirds (64 per cent) of car companies' sponsorship budget gets dedicated to sports, in comparison with spending on other sectors.[5] Our focus on car manufacturers, airlines and some fossil fuel companies is because all these sectors rank among the highest in terms of carbon emissions. The companies also largely share a business model which, being fossil fuel dependent, is in direct contradiction with the goals outlined in the Paris Climate Agreement that aims to drastically reduce carbon emissions over the next decade.

The three sectors are problematic for several reasons. Fundamentally, their products lock people into polluting, high-energy use lifestyles. All, at an industry-wide level, have lobbied against urgently needed climate action, misled and downplayed their contribution to and responsibility for the climate emergency. All, too, routinely mislead the wider public by making exaggerated environmental claims in their advertising – in other words, they 'greenwash'. Adverts for oil and gas companies, car makers and airlines disproportionately focus on typically minor environmental activities by the companies in question, while glossing over their core, polluting activities.

Oil and gas companies, for example, at a time when they were aggressively promoting themselves as leaders in the green energy transition, in reality were investing a mere 1 per cent of their budgets on clean energy. A study of spending in 2018 by the world's top 24 publicly listed oil and gas companies revealed that only 1.3 per cent of budgets totalling $260 billion went into low-carbon energy sources.[6]

Car and aviation industries are explored more in later chapters. But for years airlines lobbied for aviation emissions to be left out of international climate negotiations and the global climate plan of the industry's governing body, International Air Transport Association (IATA), literally ignores around half the climate impact of flying.[7] Car makers too have fought against action to improve air quality and cut climate pollution, with the major sports sponsor,

Toyota, being part of a lobby for measures to slow the introduction of electric cars. Other car makers have been found guilty of carbon cheating, with millions of cars fitted with devices to cheat emissions tests, making cars seem less polluting than they really are. As was pointed out at the time of revelations against car maker VW in 2015, this was only the latest in a long history of auto industry deception over harmful emissions, ranging from Ford, to GM, Honda and others.[8]

Beyond that, however, car companies have also shifted their production towards ever larger gas-guzzling vehicles – in little more than a decade the share of SUVs in new car sales has risen from around one in ten to around half of all sales. Being larger and heavier, SUVs are inefficient, using more fuel than average cars and pumping out more pollution which shifts carbon emissions in the wrong direction. Before the pandemic hit the aviation sector hard, airlines had recorded steady levels of passenger and emissions growth in recent years, despite the need to rapidly downscale the sector in the face of the climate and ecological emergency.[9]

The list of sports sponsorship deals we surveyed is full of exhaust fumes, but far from exhaustive. But still it covers a range of popular outdoor and indoor sports, international sports federations, associations, clubs, teams, leagues and other sporting events, and the list ranges across popular sports from football and basketball, to tennis, cycling, rugby, cricket, the Olympics and athletics.

All are targets for sponsors due to their far-reaching audiences. We also looked at other less mainstream sports like motorsport and sailing, which attract brands that want to connect with their more prestigious, materially aspirational images.

* * *

Why does high-carbon advertising and sponsorship of sport matter? Not only does sport reach audiences that are far bigger than those for general news and the arts, fans of sport in general, and of particular athletes, clubs and national teams develop powerful emotional bonds with them. Sponsors seek to align and benefit from this emotional connection.

Where major polluters are concerned, sport also has the same appeal that it had to tobacco companies. Its clean, youthful and healthy image offers the opportunity to 'sportswash' their polluting products.

But high-carbon companies cannot expect to keep deliberately marketing products which are driving potentially runaway, catastrophic climate destabilisation without facing any public scrutiny. The irony is that such sponsors of sport are destroying the very climate that sport, its athletes and fans rely on. Sometimes the contradictions leave an especially bitter taste.

Public transport and active travel – walking, wheeling and cycling – are the future of safer and healthier transport. For those who are able, and especially in towns and cities, cycling can not only reduce congestion and cut air pollution, but make the streets much safer for children and pedestrians and, in fact, everyone. So, for cycling of any kind to be sponsored by a fossil fuel company sends a powerfully contradictory message.

There was uproar when British Cycling announced a sponsorship deal with oil company Shell, and the ensuing scandal contributed to the departure of the sport's chief executive.[10] Another controversial petrochemical company, Ineos, has been throwing heavily marketed sponsorship deals at a range of sports, including one of the world's most high-profile cycling teams.

INEOS

A petrochemical empire promoting itself through sport, Ineos was reported in 2019 to be the world's fifth largest chemical company. It is also heavily involved in fossil fuels, fracking and significantly engaged in sports sponsorships including Ineos Team UK, the official British America's Cup sailing team worth £150 million.[11] And, gratingly, considering that cycling should be the poster child of a low-carbon transition in mobility, there is Team Ineos Grenadiers (formerly Team Sky), one of the world's highest profile professional cycling teams, prominent in races like the Tour de France, with a budget of £50 million per year.[12] Then there is OGC Nice, the

League 1 football club in France, worth an estimated 100 million euros (around £90 million).[13] Lausanne's football club, Lausanne-sport, and hockey team HCL bought for an undisclosed sum.[14] And Mercedes British Formula 1 team, with a five-year deal worth around £100 million. Ineos also sponsored the distance athlete, Eliud Kipchoge's much publicised bid to run a sub-2-hour marathon, and sponsor The Daily Mile, a programme to get children to exercise in order to improve their 'physical, social, emotional and mental health'. Unfortunate, then, that pollution from petrochemicals is particularly bad for all of those.

Ineos is a multinational petrochemical company founded and owned by Sir Jim Ratcliffe, the UK's richest man, who recently relocated to tax-free Monaco, and whose fortune is estimated at £17.5 billion, and almost doubled in value during the global coronavirus pandemic.[15]

Sir Jim's pro-nationalist sentiment, him being a staunch and vocal Brexit supporter, manifested in Ineos' recent sponsorships of prestigious UK cycling and sailing teams, as part of an apparent strategy to elevate and align the company to 'Brand Britain'. It was a move to prove controversial given subsequent business decisions (as we shall see). There is, for example, the obvious irony of environmentally friendly cycling being sponsored by a fossil fuel company, Ineos' subsidiary company Ineos Upstream Limited, which has been linked with the development of fracking in the UK.

Since the company acquired its first onshore oil in the central belt of Scotland in 2014, it committed to become a leader in the fracking industry. Ineos owns several UK petroleum exploration and development licences (PEDL) throughout the country and at offshore locations and is estimated to have development rights to around 1.2 million acres of land. A moratorium on fracking in the UK was lifted by the Conservative government using the excuse of energy security following Russia's invasion of Ukraine and volatile energy prices (this, despite energy analysts pointing out it that could not, in any meaningful way, contribute to the UK's energy security, and there being far better ways to deliver that). When they did, Ineos quickly offered to drill a test site for free to demonstrate the potential for fracking.[16] Although, amidst the turmoil of British politics, one month later another, new Conservative prime minister, Rishi Sunak, reimposed the moratorium.[17]

Hydraulic fracturing, commonly known as fracking, is a technology which involves the process of high pressure horizontal drilling as a way to extract oil and gas from rock formations. The method is particularly controversial given its high risks of water contamination with chemicals used in the process, the potential for earthquakes at the drilling sites and carbon leakage in the mining process.

Since the UK government first imposed a moratorium on fracking in 2019, Ineos has reviewed downwards the value of its assets by £63 million. However, as published in its 2019 accounts, Ineos subsidiary's strategic aim remains set on the exploration for hydrocarbons and the development of unconventional gas (fracking) in the UK. Due to fear of being greeted by protesters at the Tour de Yorkshire in 2019, the organisers of the race waited to release the location of the start of the tour until the last minute. Anti-fracking activists denounced Ineos' plans for fracking in the Yorkshire region, over which the local residents are involved in a legal battle against the company.

Ineos sponsorship of cycling can be seen as an attempt to 'sportswash' its image by attaching its name to a high-profile cycling team and a green sport.[18] Since the demise of the original off-road SUV, the Land Rover-made Defender model, a favourite of Ratcliffe's, the petrochemical magnate ordered the production of a similar vehicle – the Grenadier – to be used as an emblem for Ineos' cycling team, which was renamed for the Tour de France as the 'Ineos Grenadiers' – a cycling team named after a 4x4 SUV.[19] Ironically, the supposedly British-based engine will now be made in France at Hambach rather than Bridgend in South Wales, the originally destined site, a manufacturing location which the company bought from car manufacturer Mercedes-Benz.[20] This move came as a shock and gave rise to accusations of hypocrisy given Ratcliffe's support for Britain leaving the EU.[21]

In 2010, Ineos moved its headquarters to the town of Rolle in Switzerland for tax purposes. The town currently hosts Ineos

Europe AG, after the company moved its main headquarters back to London in 2016 following the government's cuts in corporate tax. Seven years later, the company's CEO, David Thompson, made a name for himself in the Swiss area by taking over the sponsorship of both famous local football and hockey clubs, FC Lausanne-Sport and HC Lausanne.[22]

The company has great ambitions for the football club and hopes to have it qualify for European championships. Contrary to its tainted image in the UK, the company's popularity remains untouched in the local area.[23]

This partnership reveals its particular importance for the petrochemical company, which it is using as an entry point into the European market.

In 2019, Ineos invested 3 billion euros to expand its chemical production facilities in the port of Antwerp in Belgium, the largest chemical cluster in Europe.[24] This project builds upon the company's business model of importing by-products. To finance this project, the petrochemical company called on institutional investors, such as the banks BNP Paribas, ING and Deutsche Bank, as well as on taxpayers' money.

The presence of Ineos Grenadiers' vehicles in Belgium was recently met with backlash from activists, who spray painted 'Ineos will fall' on the cycling team's vehicles in protest against the company's chemical expansion in the Antwerp port.[25]

Besides their impact on the climate, transport emissions are particularly damaging for health reasons. Vehicles emit large amounts of pollutants which, especially in urban areas, come into direct contact with people – pedestrians, cyclists and, of course, drivers too. Of concern are mainly nitrogen oxides (70 per cent) and particulate matter (30 per cent) according to the European Environment Agency. Despite these being subject to law governed by European legislation, populations living in urban areas still breathe air that is above safe limits.[26]

It is estimated that an adult breathes around 15,000 litres of air per day, and if the air is polluted it can cause damage to the lungs and airways, enter the bloodstream and damage internal organs.

Full understanding of the consequences is evolving, with both the range and scale of health impacts constantly being revealed as larger than previously thought.[27]

For the World Health Organization (WHO), out of their estimated 8 million premature deaths around the world caused by air pollution, half of these are due to ambient air pollution, the rest being household-related pollution (heating and cooking, in other words). Around 500,000 lung cancer deaths and 1.6 million cases of COPD (chronic obstructive pulmonary disease) are caused by air pollution.

Added to this, air pollution also contributes to an estimated 19 per cent of all cardiovascular deaths and 21 per cent of all stroke deaths.[28] But more recent research from three British universities and Harvard looked again at the impact from fossil fuel-related air pollution alone, using a more sophisticated analysis. It found that more than 8 million deaths could be attributed alone to just fossil fuel air pollution, accounting for nearly one in five of all premature deaths globally.[29]

Medical research is constantly revealing ever wider impacts from air pollution, including on mental health. Following the pandemic outbreak in 2020, many studies have established that living in areas with poor air quality increases the severity of Covid symptoms.

Where air pollution is concerned, these impacts disproportionately affect people of ethnic and racial minorities and lower income and socio-economic status. Crucially, the latest scientific research concludes that among people living in heavily populated areas (in this case US cities) where they are exposed to high levels of fine particulate matter, they also had an 11 per cent higher chance of dying from Covid-19.[30]

The recent results have a similar significance to the relationship established between smoking and diseases.

TOYOTA: A RECORD-BREAKING SPONSOR

Oscar Wilde once quipped that if you give someone a reputation for early rising they can stay in bed till noon without it being noticed.

Toyota's early entry into the hybrid vehicle market – which at the time seemed progressive – gave them a green reputation.

Now, however, hybrids are seen as much more problematic, having made exaggerated claims of reducing emissions, and being seen as a way for car makers to prolong the manufacture of polluting, internal combustion engined cars. And, since Toyota's early move into hybrids, it has stayed in bed until long past noon in terms of shifting production to fully electric vehicles, or BEVs.

Worse, it has lobbied to obstruct industry-wide measures to accelerate the shift to BEVs. In a single year, 2021, Toyota sold 10,383,548 internal combustion engined (ICE) vehicles that rely on fossil fuels. Over their lifetime, these vehicles will emit over 700 million tonnes of CO_2, the equivalent of 190 coal-fired power stations running for an entire year. From 2022 to 2040, Toyota is planning to sell 110 million ICE vehicles. The total emissions from these vehicles over their lifetime will be in the region of 7.4 billion tonnes of CO_2, which is equivalent to running more than 2,000 coal-fired power plants for an entire year.[31]

Toyota, headquartered in Toyota City in Aichi, Japan, is the largest car manufacturer in the world.[32] It owns several other car brands including Daihatsu, Hino and Lexus, and has large stakes in several other car and motorbike companies, including Subaru Corporation (20 per cent), Mazda (5.1 per cent), Suzuki (4.9 per cent), Isuzu (4.6 per cent) and Yamaha Motor Corporation (3.8 per cent).[33]

Toyota has a global reach. North America remains one of its largest markets, with nearly 2.7 million vehicles sold there in 2021.[34] In the same year, just over 1 million Toyota cars were sold in Europe and over 3 million in Asia, with the lion's share in China.[35] The growth in Toyota's sales in 2021 was not equal around the world. In some markets, like Germany and Thailand, sales fell.[36]

But Toyota recorded a huge increase in sales in other emerging markets, such as a 65 per cent rise in Pakistan, almost 59 per cent in Indonesia and over 30 per cent in South Africa.[37] All these countries are grappling with devastating climate impacts ranging from flooding to droughts, as well as from high levels of air pollution.

The company, though, has long viewed sport as an ideal billboard on which to market its vehicles. Among the ten biggest car brands of 2021, Toyota had the largest number of sponsorship deals by a car brand globally, including the largest number of active sports sponsorship deals, and the highest market share (10.5 per cent). At the end of 2022, including both past and present deals, at least 23 different sports were covered by the car maker's sponsorship deals which numbered over 90 active deals.

Toyota's 2015–23 partnership with the Australian Football League is estimated to be worth $18.5 million (Australian dollars) per year.[38] Toyota's deal with the Olympics for the period 2017–24 is worth an estimated $835 million (US dollars) – four times the cost of previous partnerships.

The Japanese car maker is the first car company to join the International Olympics Committee TOP sponsorship programme which gives companies exclusive worldwide marketing rights.[39]

Toyota USA Motor Sales is the most active car maker to sponsor sporting events in the USA: out of all sports sponsorship deals with an automotive company, just over a third of them (34 per cent) are signed with Toyota.[40]

THE IMPACT OF CLIMATIC UPHEAVAL ON SPORT

The greatest irony is that despite receiving sponsorship from companies that are directly causing havoc to the climate, global sports are particularly vulnerable to the present and future impacts of climate breakdown.

Research from the University of Waterloo found that by 2050, around half of the cities which have hosted the Winter Olympics would be potentially too warm for outdoor snow sports.[41]

To raise awareness about the issue, Finland's coldest town of Salla placed a bid to host the 2032 Summer Olympics.[42] In a public announcement, the mayor of Salla said: 'If we stand back and do nothing, letting global warming prevail, we will lose our identity, and the town we love – as well as many others around the world – will cease to exist as we know it.'[43]

Australia, in particular, is a country prone to the devastating impacts of climate change on its sports sector, whose economic value is estimated at AUS$13 billion.[44] With the mercury set to reach over 35°C for several days in the year by the end of this century, this will render sports conditions increasingly risky for athletes.

The 2014 Australian Open is often cited as a particularly striking example of the numerous challenges posed by extreme heat in sports. With temperatures reaching above 40°C on the tennis court, conditions for players were deemed 'inhumane' forcing some to abandon their games and around 3,000 spectators to leave matches halfway through.[45]

Extreme temperatures are particularly dangerous for human bodies, with increased risk of organ failure above 38°C and cases of death past the 40°C threshold. Besides tennis, sports like football, running and marathons are especially vulnerable to high temperatures and humidity.[46]

Following record-breaking heat for the season, at the 2015 Los Angeles Marathon more than 30 runners had to be admitted to hospital.[47] A study into high school football found a threefold increase in heat-related deaths between 1994 and 2009 compared to the previous 15 years.[48]

Droughts, on the other hand, are particularly damaging for sports like cricket and golf. In 2016, 13 Indian Premier League games had to be moved away from Maharashtra which was experiencing the most severe droughts in a century. Droughts also affect river sports, such as canoeing, due to lower water flows.[49]

Conversely, flooding is a serious issue for countries playing sports in more temperate regions.[50] Cricket grounds and football stadiums in England, the Netherlands, the USA and Canada are especially at risk of flooding.

Research predicts that out of England's football league teams, almost one in four, 23 in total, will have experienced partial or full flooding by 2050.[51] In addition, the increased frequency of climate hazards such as storms, hurricanes or forest fires will inevitably hinder the capacity to play outdoor sports.

A groundbreaking report by the Rapid Transition Alliance, *Playing against the Clock*, into the links between the climate crisis and global sports, estimates the sector's overall carbon footprint to be in a range which, at the low end, would be equivalent to that of a nation like Bolivia, and at the higher end equal to the emissions of countries like Spain or Poland.[52]

The two largest events in global sports, the Olympics and the World Cup, release as much as 7–8 million tonnes of greenhouse gases into the atmosphere. And, based on some estimates, the average annual emissions for a football club amount to around 10,000 tonnes of greenhouse gases per year.[53]

NORMALISING POLLUTING LIFESTYLES: HOW SPORT SETS SOCIAL NORMS AND WHY IT MATTERS

There is a strong economic rationale for corporations to engage in sports sponsorships. According to several studies, the motives behind corporate sports sponsorship are principally for commercial advantage. Companies use it as a marketing tool to create a positive public image for themselves or increase their TV coverage. Philanthropic motivations are rarely, if ever, mentioned.[54]

Other research established that consumer purchases are positively influenced by the belief that a company's brand is involved in sports sponsorship.[55] The relationship was found to be even stronger for customers who had purchased the brand on a prior occasion.[56]

Furthermore, research into the impact of sports sponsorship also shows that it influences consumer behaviour by creating a positive association between the brand and the spectator's sports team.[57]

To use the jargon of the industry, sponsorship can also have an effect on consumers' 'attitudinal cognitive component' by modifying their perception of the brand.[58] As described in the example below, corporate brands justify their sponsorship on the basis of creating strong and trustworthy relationships with their customer base.

General Motors, former sponsor of the Olympic Committee, the Women's National Basketball Association, the PGA Tour and PGA of America, put it like this:

> We are now trying to be more focused on what we are doing, and we want to go much deeper with each relationship. Instead of going a mile wide and an inch deep, we want to go a mile deep and an inch wide. It's all about creating stronger connection points with consumers based on their passions.[59]

Companies go to great lengths to position themselves more broadly in line with their sports sponsorship.

In relation to its partnership with the Tokyo Olympics and Paralympics Games, Toyota GB's marketing director argues: 'Toyota wants to make sustainable mobility accessible for everyone, regardless of age or physical ability. This mission to deliver ever-better mobility for all is at the heart of our brand and resonates with the way in which athletes can inspire people to go further and realise their dreams.'

'Tokyo 2020 is an extremely important and exciting time for the Toyota brand as our business begins its evolution from an automotive manufacturer to a mobility company and we look forward to bringing this to life through this partnership and beyond.' Toyota may talk the talk, but it is still filling its tanks with petrol and diesel. Keep in mind at this point that the car company is still planning to sell around 110 million ICE vehicles by 2040 with lifetime emissions estimated to be in the region of 7.4 billion tonnes of CO_2.

On other occasions, companies give more practical reasons behind their associations. The airline Emirates, for example, justifies its sponsorship of the French football club Olympique Lyonnais through connecting flights between the cities of Dubai and Lyon:[60]

> With Olympique Lyonnais, we've found a partner that mirrors our 'Fly Better' brand promise of striving to achieve the highest levels of success, and there is already a connection between Lyon and Dubai with Emirates' daily flights between both cities.[61]

Besides economic value, sports play a significant social role. In many countries, sports are a dominant feature of the cultural landscape in terms of participation, spectacles to watch, elite athletes with celebrity status, and sheer volume of media coverage. In India, a 2015 cricket match between India and Pakistan was watched by a billion people – approximately a seventh of the world's population.[62] International events like the Olympics Games can attract as much as half of the world's population (3.6 billion people) as recorded during the 2012 Games.

These facts demonstrate the incredibly powerful reach of sponsorship. More than anything, partnerships between corporations and sports create a 'cultural licence to operate'. In other words, it grants companies a certain sense of credibility and normality in society, simply by having them associate their name with established sports teams and events.

If a company has, or is vulnerable to, reputational problems, the term 'sportswash' has been used to describe the potential benefits to companies.

This is something which the advertising industry itself is becoming aware of at the highest levels. Addressing an Advertising Association webinar, AdNet Zero, in December 2020, focused on the industry's track record on climate, Charles Ogilvie, strategy director for the COP26 climate conference for the UK government, said:

> From the advertising sector's perspective, sometimes it will be better for your client not to turn up [to the COP26 climate talks]. And I say that quite clearly. It will do them no good because they will be pilloried because their attempt to greenwash themselves will ultimately not succeed and will be highlighted by protesters and commentators and potentially even politicians. So it will be a bad move for them and it will be bad for us because it will become noise.[63]

THE 'SOFT POWER' OF SPORTS SPONSORSHIP

In contrast with car companies, or airlines whose involvement in sports sponsorship is focused on brand and reputation building, and increasing sales, fossil fuel companies such as Gazprom or Ineos may typically have additional aims. Their sponsorship of large-scale sports events or world-renowned sports teams can be leveraged for geopolitical influence.

Amnesty International first talked of a case of 'sportswashing' in relation to the deal secured between the UAE capital, Abu Dhabi, and football club Manchester City.[64] Amnesty sees ties with big sports as an attempt for the country to improve its image in relation to human rights abuses.[65] Similarly, companies like Gazprom or Ineos may use the 'soft power' of sports to position themselves in the energy field.

GAZPROM – FOSSIL-FUELLING EUROPEAN FOOTBALL[66]

Gazprom, the world's largest gas producer and state-owned Russian company, has been reported to use its sports sponsorship contracts as a way of strengthening its political influence. Before Russia invaded Ukraine, Gazprom sponsored four European football clubs: Zenit Saint Petersburg, Schalke 04, Chelsea and Red Star Belgrade, as well as the UEFA champions League (a nine-year partnership running from 2012 to 2021) and the prestigious FIFA World Cup. The numbers are big. Gazprom signed a $90 million (€71.7 million) four-year sponsorship deal with FIFA that included the 2018 World Cup. The previous sponsor of the UK club, Chelsea, the Russion oil company Sibneft, sold its shares in the football club to Gazprom for an estimated $13 billion (€10.4 billion). In 2010, the company secured a five-year deal with Serbia's football club, Red Star Belgrade, worth around $19 million (€15.2 million).[67] And when the latest deal with German football club Schalke was extended until 2022, it was estimated to be worth €20 million.[68]

It took a war for Gazprom to be dropped, somewhat reluctantly, from sponsoring competitions outside Russia.

In contrast with other sports, football sponsorships tend to include a legal right to use a property's name and give access to corporate hospitality facilities inside stadiums. These are very useful tools for gathering geopolitical influence in terms of raised profile and being prestigious venues, convenient and attractive for holding meetings, formal and informal, with decision-makers. In 2006, the company's logo appeared on the German football team FC Schalke 04 – also referred to as 'the Miners' given their location in the coal heartlands of Germany.

This deal followed the approval of a gas pipeline a year earlier – the NordStream 1 – directly connecting Russia and Germany. Before the war in Ukraine, Gazprom provided 35 per cent of Germany's gas supplies, which makes the country its largest foreign market.

In 2006, the company had also signed a deal with the Russian football team Zenit. Such deals give good publicity to the company, which even arranged for the team to play on one of its gas platforms.[69] The deal also helped build ties between Russia and Germany whose teams Zenit and FC Schalke 04 played in a friendly match in 2007. In 2010, Gazprom took over sponsorship of Serbia's team Red Belgrade and simultaneously secured a pipeline project in the area – the South Stream – which was later cancelled in 2014.[70]

Greenpeace staged a protest at a match between football teams FC Basel and FC Schalke 04 to expose Gazprom's role in exploiting the Arctic.[71] The company's influence remained strong, however, and it has been adding its name to major football events, such as the 2018 World Cup and the UEFA Euro 2020 championships.

THE OIL STAIN ON THE ARTS

High-carbon sponsorship of the arts, by fossil fuel companies in particular, is something that grassroots groups have campaigned for many years to remove. Organisations like Culture Unstained, BP or not BP, or Liberate Tate have criticised the ties fostered by fossil fuel lobbyists with high-profile cultural and arts institutions, such as the National Portrait Galleries, the Tate Modern or the Royal Shakespeare Company.

Thanks to successful campaigns led by these groups and increasing public pressure, some sponsors have had their names removed from these arts institutions. According to campaigners, victories are far from merely symbolic. In fact, they argue, these show that:

> through sustained pressure, public rejections of BP and other fossil fuel firms have become an effective tool for challenging what is known as the industry's social licence to operate. Warm words have done little to end the industry's pursuit of fossil fuels, but cutting cultural and financial ties is now changing the discursive space.[72]

These groups, and other arts organisations, can hardly pretend any more to be merely 'neutral' institutions in the face of fossil fuel sponsors, who have used their 'progressive, educational agenda' as a tool to 'artwash their image'.[73] Similar reasoning can be applied to the case for ending sports sponsorship by high-carbon companies, whose ties with sports provide them with an increasingly powerful platform to curate their public image.

In 2023, there was a glimpse of how history's dial turns on what is considered acceptable, and about how principled campaigners, who one day may be considered a nuisance, reviled or even arrested, assaulted or worse, may in time even be thanked by the targets of their protest. Such was the case in 2023 when Francis Morris, the director of the Tate Modern, once sponsored by oil company BP, said this in a talk at the Whitechapel Gallery:

> Tate declared a climate emergency in July 2019 and we did so with the encouragement of the group Culture Declares who came into our Turbine Hall. And that had been preceded by many years of extraordinary interventions in our Turbine Hall and on our landscape by Liberate Tate. Liberate Tate were the activist organisation who really took (oil company) BP out of the institution of Tate. They did so in the most incredibly compelling way ... when I talk about the Turbine Hall as a civic space now – I mean not only showing commissions of artists that we have brought into the space but also the equally extraordinary activist works that were just placed there.[74]

THE CASE FOR ENDING HIGH-CARBON SPONSORSHIP OF SPORT

Awareness of the scale of the climate crisis has struck the world of global sports more recently. For example, a group of 300 Olympic and Paralympic athletes wrote a letter to the UK government pressing it to consider a specifically green recovery from the global pandemic.[75] The letter was released through the Champions for Earth initiative. Among its founding members, environmental lawyer and Team GB rower Melissa Wilson said:

> Each of the Champions For Earth athletes have experienced what can be achieved by individuals coming together in a team – creating something that is greater than the sum of its parts. Climate change is the issue, more than any other, that needs that kind of collaboration – because it affects every single one of us.[76]

In a similar move, a group of nearly 200 mostly winter sports athletes wrote to their governing body in February 2023, the International Ski and Snowboard Federation (FIS), sounding the alarm over the threat to winter sports from global heating, and calling for changes in the organisation and structure of the sport to lower its own impact, such as reducing the need for international flights between poorly scheduled international events that create the need for unnecessary travel.[77]

The issue of sports sponsorship also logically falls under the remit of responsible consumption listed under Principle 4 of the UN Sports for Climate Action Framework. At a more institutional level, the UN recently launched a platform for implementing the recommendations from the Paris Climate Agreement in the sector. Organisations which sign up to the Sport for Climate Action Initiative must commit to a set of five principles and demonstrate their progress towards meeting each of these targets as part of their overall engagement.[78]

Principle 4, in particular, states that sports organisations and events organisers should have sustainable procurement policies in place in order to promote sustainable and responsible consumption. Besides the obvious climate impacts generated from transport, building materials or energy provisions, the case of sports sponsorship also logically falls under the remit of responsible consumption.[79]

Many high-carbon companies controversially sign onto scientifically dubious carbon offsetting programmes, while keeping their core business practices largely unchanged. It is equally questionable for sports organisations to claim climate neutrality while accepting money from companies which are directly undermining their climate commitments.

If global sport is to take the issue of climate breakdown seriously, it must be consistent and coherent and review its partnerships with organisations whose practices go against their efforts to safeguard the future of our planet. In 2019, Paris mayor Anne Hidalgo set a precedent in the field by denying the fossil fuel giant Total access to sponsor the 2024 Olympics, stating that 'it would be very difficult for our fellow citizens to identify with companies whose activity would have a big impact on the environment, based in particular on the massive use of carbon-based energies'.[80]

WHY OFFSETTING IS NOT A POLLUTER'S 'GET OUT OF JAIL FREE' CARD

So while we are on the subject, let's just think about carbon offsetting for a moment. This is now widely used by companies who sign up to greenhouse gas emission reduction programmes.

The practice allows companies to keep on emitting a certain amount of carbon emissions and be given 'carbon credits' in exchange for investing in carbon reduction projects, such as reforestation schemes, in other countries – often in the Global South.

Among global sports institutions, FIFA is a high-profile organisation relying on carbon offsetting to live up to its climate promises. By pledging to become carbon neutral by 2050, the international football federation sets a precedent for others in football to use offsetting to attempt to achieve environmental objectives.

But carbon offsetting schemes have been criticised on several grounds, for their highly 'limited-to-nil' impact on reducing emissions, to outright negative impacts when there are devastating consequences on local communities where some projects take place. The schemes grant large polluting companies, like fossil fuel firms or airlines, the licence to keep polluting while engaging in projects that supposedly compensate for carbon pollution that has already happened. But these schemes often lack scientific credibility while functioning in support of greenwashing and PR strategies for reputation management.[81] A 2016 study for the European Commission found that the UN's offsetting mechanism, which FIFA signed onto, had significant flaws and that only 7 per cent of emissions reductions credits were of substantial and measurable value.[82] More recent research into the world's largest supplier of rainforest carbon offsets found over 90 per cent of these to be 'worthless'.[83] In fact the charge sheet against offsetting is long, and this is crucial, because offsetting is the first solution typically reached for by organisations, companies and individuals once the need for climate action has been accepted.

The basic problem is quite simple, offsetting doesn't work. In its current form scientific evidence suggests that offsetting simply doesn't deliver on its promises. It can cause real harm, both in terms of direct project failures, but also in terms of the impact of offsetting projects on local communities, economies and the natural world. It is also a form of carbon laundering. Attempting to offset stable stores of fossil carbon with unstable stores like trees, which face multiple threats in our warming world, masks the true climate impact of the original emissions. Offsetting claims can be gamed, through accounting tricks and murky carbon markets, offsets can be misallocated on a mass scale, which often means there is no reduction in overall emissions. It also provides an excuse to continue polluting. Offsetting can justify and legitimise the status quo, allowing organisations to continue polluting while claiming leadership and progress on sustainable and environmental issues. Ultimately, it inhibits real change. The cost and convenience of offsetting means that the more challenging, structural decisions required to address sport's climate impact may be delayed. As an approach, offsetting restricts our thinking and ambition around how sport can drive climate solutions and as part of a thriving planet.[84]

For many of these reasons, scepticism about offsetting claims in adverts spread to such a degree that in May 2023, the European Parliament passed a bill that prohibits companies from advertising carbon offsets because 'this practice wrongly leads consumers to consider related products or services as safe for the environment'.[85]

The multiple health, social and local economic benefits of localised, low-carbon sport has not only many proven benefits for physical and mental health but it can also play an important role in the local economy and community – by bringing people together, encouraging more localised spending, and forging relationships across different cultures and socio-economic backgrounds.

Local sports clubs can also set an example for fostering meaningful climate action and contributing to local, low-carbon rapid transition. Organisations like Forest Green Rovers in the west of England, the first UN certified zero carbon football club, are taking the lead on tackling carbon emissions.

Forest Green Rovers runs on 100 per cent renewable energy, serves strictly vegan food to its players and fans, installed rainwater recycling, a solar-powered lawn mower and electric vehicle charging points on its grounds.[86] More recent developments in the field include clubs like Real Betis in Seville which has committed to being 'climate neutral'. German football clubs Mainz SC and FC Freiburg, on the other hand, have been pioneering sustainability practices for more than a decade now, by installing green waste management schemes and powering their stadiums with renewable energy.

These cases illustrate the multiple positive impacts that local sports clubs can have on their local community and the wider environment.[87] With sports fans looking up to their favourite local sports teams and players as role models, such initiatives can act as a powerful tool to foster broader positive behavioural change.[88]

Reforms in the field of sports, such as those initiated by the UN Sports for Climate Action Framework, are welcome and necessary to meet international climate targets. However, it is imperative that calls for climate action amount to more than mere publicity campaigns from sports clubs and events to promote an environmentally responsible image.

For sports publicly advertising their green credentials, such as sailing – where the highest profile outfit is Ineos Team UK's boat *Britannia* – or cycling, which should be a poster child for clean, affordable, enjoyable transport, where one of the highest profile international teams is not only sponsored by the same petrochemical company, but used to promote its SUV, the Grenadier, it is not only inconsistent but self-defeating for sport to be a billboard for advertising and sponsorship from companies who fuel the climate emergency.

As Lizzy Yarnold, Britain's most successful Winter Olympian, said at the launch of *The Snow Thieves*, the Badvertising report into high-carbon sponsors of winter sports:[89]

> At their best, winter sports are a celebration of people enjoying some of the most awesome landscapes on Earth. But the impact of climate pollution is now melting the snow and ice which these sports depend on. Having high-carbon sponsors is like winter sport nailing the lid on its own coffin, and it needs to stop.[90]

5
How Big Car Persuaded Us to Buy Big Cars

> Which auto-driver has not felt the temptation, in the power of the motor, to run over the vermin of the street – passers-by, children, bicyclists?
>
> —Theodor Adorno, *Minima Moralia*, 1951

> Put the world at the mercy of your whims.
>
> —Isuzu advert ad slogan for SUVs, 1998

There was a time in the early 1980s when the technology was available to make cars smaller, more efficient and far less polluting. It would have been good for people's health, the neighbourhoods where they live, the climate and life on Earth. Cleaner air, less congestion, safer streets, more pleasant towns and cities. But that's not what happened. Instead, the opposite came to pass. Cars got bigger, more polluting and more lethal on city streets. How? Car makers first developed and then heavily marketed the sports utility vehicle, or SUV. We live with the consequences still; as of 2023, SUVs make up nearly half of new car sales around the world, a share which rises further with each passing year.[1]

* * *

Car adverts often display images of exotic, wild or rural locations surrounded by plenty of space, and this is especially true for adverts for SUVs. The ads promise adventure, escape and the open road; but the reality in many world cities today is ubiquitous traffic jams and soaring, illegal levels of air pollution.[2] The paradox of car adverts which depict roads that are always devoid of other cars is that the more successful these visions are, the more outrageous the lie they are telling becomes.

Through this seductive promise, over the past 80 years, the car has become culturally dominant, not merely as a mode of transport but as an active presence in the public realm and, crucially, as the pre-eminent positional good through which to signal social status to others. Car advertising has played a key role in the ascendance of car ownership as the ultimate status symbol, setting into motion cultural forces which would eventually lead to 'keeping up with the Jones's' escalating into a full-scale arms race on our roads.

Recent years have seen vehicle manufacturers move the bulk of their advertising spend to promoting their bigger, more polluting SUV ranges rather than traditional family cars. In 2018, the car maker Ford's dealers reportedly spent 66 per cent of their advertising budget promoting SUVs and light trucks in the USA – up from 42 per cent in 2016. In the same period the car maker itself reportedly shifted its ad spending from a 50/50 split on cars and trucks/SUVs, to spending 85 per cent of its marketing budget on trucks/SUVs.[3] For decades the Ford Fiesta was the UK's most popular car, but it will be discontinued this year – not, as the industry would have us believe, because people no longer want to buy it – but because Ford no longer want to sell it. The short answer to the question of why a car manufacturer like Ford would choose to stop selling its most popular product is this: profit. SUVs are enormously more profitable than smaller cars like the Fiesta, costing a similar amount in labour and components to produce, but commanding a far higher sales price; the early SUVs provided a 25 per cent profit, compared to just 5 per cent on ordinary cars: Ford were able to buy Volvo and Land Rover with their SUV profits by 1999.[4] Shifting production and marketing from less profitable product lines to more profitable ones is the only economically rational thing for a car-making business to do – regardless of the wider implications for society.

Before we go any further, readers may be wondering what exactly makes a car a 'sports utility vehicle'. The answer is surprisingly vague. There is no hard and fast definition for what a SUV is, but the common theme that unites vehicles which are marketed as SUVs is that they all incorporate design elements of off-road vehicles. In some cases (e.g., some of Land Rover's products) they are bona fide off-road vehicles which can handle rough terrain thanks to four-wheel drive, increased ground clearance, low range transmissions and differential locks. In most cases SUVs are not at all suited to off-road driving and are essentially no different to conventional cars other than having taller, heavier bodies, an elevated driving position and 'rugged' design cues – all sold at a premium price. The increased size, weight and drag of these vehicles leads to higher fuel consumption and CO_2 emissions compared with conventional vehicles, as well as increased risk to pedestrians and other road users.

The industry's drive to steer us to buy these larger, dirtier vehicles has been so effective that it is now wrecking progress to our climate change targets; in 2016 the average CO_2 emissions of a new car sold in both the UK and the EU stopped falling and began to rise. This reversal is entirely thanks to the exploding market share of new SUVs, energy-hungry vehicles which typically produce around 25 per cent more CO_2 emissions than a medium-sized car.[5] So far, people have been switching to more polluting SUVs much faster than they are switching to less polluting electric vehicles, more than cancelling out the emissions saved. Transport is now the UK's and Europe's biggest source of carbon emissions, contributing over a quarter of the EU's total CO_2 emissions in 2018, with cars and vans representing more than two thirds of these.[6]

As well as being dirtier, oversized SUVs take up far more precious urban space than conventional cars. In 2019, over 150,000 new cars were sold in the UK that are too large to fit in a standard parking space.[7] Yet instead of enforcing parking controls, many authorities have simply shrugged and unquestioningly begun repainting the lines to accommodate this bloat. These enormous vehicles, ostensibly built for rural, off-road work, are mostly being bought by people in towns and cities. The part of the UK where the largest and most powerful four-wheel-drive SUVs – so-called 'Chelsea Tractors' – are most popular is indeed the inner London borough of Kensington & Chelsea, where there are no opportunities for off-roading.[8] Public awareness of this incongruity grew during the coronavirus pandemic, when we needed as much space on our streets as possible for pedestrians and cyclists to get around and commute to work safely.

Many urban centres are struggling to provide functional transit systems based on majority public transport or active travel because our infrastructure is drowning in unmanageable volumes of private cars. Excess traffic makes roads too dangerous to cycle and mires buses in congestion, feeding a destructive spiral which ushers ever more people into the gilded cage of car dependency. The growth in not just the number but the physical size of private motor cars has emerged as a key barrier to creating liveable cities and meeting vital climate change goals.

The question explored over the rest of this chapter is exactly how these American-themed oversized cars came to dominate the global automotive market. Why are so many otherwise sensible people now buying SUVs, despite their inconvenient scale mismatch with narrow city streets, poor fuel efficiency, extra pollution and worrying safety record? How have manufacturers persuaded nearly a whole generation of car owners that they need a two-tonne truck to make short urban shopping trips?

* * *

To find the origins of the seemingly irresistible rise of SUVs, we need to go back more than half a century, to the dawn of mass motoring in the USA in the 1950s, heralded by the opening of the world's first – and possibly only – drive-thru wedding chapel in Las Vegas in 1951. President Dwight Eisenhower endorsed the interstate highway system five years later, in 1956.

The following decade saw Jack Kerouac's driving novel *On the Road*, plus the opening of Route 66 from the Atlantic to Pacific coasts, and slick road movies like *Bullitt.* For the 1950s generation, a car meant freedom from parental control, making out on the back seats at drive-in movies. It meant sexual freedom. You can understand its potency.

By the 1960s, thanks to the arrival in the US market of European and Japanese cars, the market had begun to segment – you could choose fast and cool like James Bond in his Jaguar or small and trendy in a Morris Mini-Minor, or tuning in and dropping out with the hippy image of a VW camper van.

But all these different approaches to marketing motor cars came from similar roots: they all offered the promise of escape. And nothing looked like escape quite like the original Model T Ford first produced in 1908, driving up mountains or hiking around the countryside with Ford, Edison and Firestone.[9] In fact, when production of the Model T finally stopped in 1927, there may have been vastly fewer cars than today, but as many as 68 per cent of the world's cars were Model Ts.

So much car advertising between the wars on both sides of the Atlantic was designed to imply effortless superiority. As in 'Let it pass … it's an Alvis' (UK, 1920s).

Or the first advert to use the trick that SUV advertisers would turn to in the 1990s – showing the car by looking up at its radiator, a kind of abject prostration before the motor, as in 'The newest car in the world', looking up beyond the car at the modernist Guggenheim Museum (Cadillac, 1939).[10]

By the 1940s, car adverts got tougher, things were not quite as effortless any more. 'Pay off for Pearl Harbor', said an advert for Cadillac with a picture of a bomber overhead – another trick that SUV advertisers would learn from – saying how tough it was out there and that your car could be your best defence (Cadillac, 1944).

This toughness was exemplified by one car in particular, the American Jeep. The Jeep was a quarter-ton truck from the American Bantam Car Company. Even before the war came to the USA, Bantam found they could not keep up with demand, so in 1940, they signed contract licences over to Ford and Willys-Overland. Willys produced 350,000 Jeeps during the Second World War and Ford 280,000. In 1945, Ford gave up production of the Jeep and Willys went on to develop a four-wheel-drive passenger version they called the Jester.

Four decades later, the Jeep would go on to provide the archetype for the modern SUV craze. But first, a series of quirks in the development of American motor manufacturing regulations had to take place to create the conditions for such an irrational trend to be able to take off.[11]

The first twist emerged as part of a post-war spat between the USA and the European Economic Community, as it was then, which came to a head in the mid-1960s. In the days before chlorinated chicken, in the 1950s, American chickens were imported into Europe in great numbers, and especially into Germany. The forerunners of the EU wanted to develop their own agricultural sector, so they imposed a tax on imported chickens. The Lyndon Johnson administration responded with a tax on imported pick-up trucks. In effect, this locked foreign competitors out of the US off-road market for a generation, until the turn of this century. That meant that, for nearly four decades, the off-road vehicle market in the United States was exclusively American-made.

Also important was a mistake made by the chairman of transport safety in the Carter administration ten years later. Fuel economy regulations introduced in response to oil shocks in the 1970s were applied to all passenger cars – but not, thanks to successful industry lobbying – to trucks. Further successful lobbying saw 'trucks' defined vaguely as 'an automobile capable of off-highway operation'. This meant that there were for many decades no fuel economy rules when your 'truck' was over 8,500 pounds in weight – including passengers on all the seats and cargo. New 'gas-guzzler' taxes and standards applied to smaller cars made them more expensive to manufacture, while much bigger passenger cars that could be classed as trucks faced none of these extra costs. Manufacturers realised that giant cars could become their cash cows. But first they had to work out how to get customers to buy vehicles that were far bigger than they actually needed.

In 1984, the first SUV successfully promoted in the massive American market was the Jeep Cherokee, which began recording high sales that year, as the term SUV was beginning to emerge in the motoring press. That came to the attention of a Ford executive called Bob Lutz, who had been exiled within the company to trucks, then a corporate backwater representing a small fraction of the market. He started developing ideas for a new kind of mainstream vehicle for general driving use that was as closely related to a truck as to a normal family saloon.

Two years later, Lutz had made his case to the company board and was recruiting his team. The Ford board had been sceptical, as well they might be: why would anyone buy a car based on a light truck which could only manage 20.5 miles per gallon (mpg) – especially when they could spend less on a more traditional car which could go 27.5 mpg or more? The answer, startlingly enough, came from the UK.

The Range Rover went on sale in the UK in 1970, as a two-door, larger Land Rover. It was not originally designed as a luxury car, far from it. In fact, its utilitarian plastic dashboard was intended to be washed with a hose. But a few of them made it to the USA via the grey market, and one of these was imported to Detroit by Edsel Ford II, grandson of the company founder.

Range Rover branding sought to put across a combination of safety and wartime strength – plus a hint of testosterone-fuelled suggestiveness becoming common in other car marketing: 'It pulls beautifully' (overtly sexual innuendo for the Triumph Spitfire, 1979).[12] Range Rover often also played on the British penchant for outright snobbery: 'There's only one car for the double-barrelled' (grouse-shooting with Range Rover, 1981).[13] Range Rovers were meant for the type of person who is both posh and armed: 'In a country that has 184 rainy days, 51 freezing days and 25 foggy days every year, no car can ever be too safe' (Rover, 1973).[14]

Edsel backed Lutz to build the Ford Explorer because he wanted a US equivalent from Ford. That was how the decision was made. Lutz recruited the top Ford engineer Steven Ross and, by 1986, the team was in place in a big room covered with photos of contemporary films like *Top Gun*, *Rambo* and *Rocky IV* – signalling the next phase in the SUV's marketing journey.

By then a huge amount of market research had been carried out on the unsuspecting Cherokee owners, who, as it turned out, did not want mini-vans, which they described as 'mum mobiles'. A typical Cherokee owner was a man with a family, who wanted to send a different message via their car than a 'docile, family' one. He wanted something to make him feel younger again and more carefree.

The psychological needs of insecure men going through their midlife crises were a perfect match for the SUV's rugged public image: these buyers never needed off-road 4x4 capability for any actual journeys, but they still wanted to have it.

Marketing for the Ford Explorer, which Lutz's team went on to design, emphasised toughness and a sense of safety, especially in city streets, which they portrayed as unpredictable and threatening, just as SUV advertisers have since: adverts for the Chevy Blazer claimed that it provided 'a little bit of security in an insecure world'. One of the top motor design consultants in the car-making city of Detroit, Clotaire Rapaille, who had a background in Freudian and Jungian psychology, said that 'SUV buyers want to be able to take on street gangs with their vehicles and run them down.'

One other key development that fed into the creation of the SUV market was the first Gulf War. Introduced in 1979, the HumVee was a military transport and assault vehicle used by the US army that had featured prominently in news coverage of the 'Desert Storm' operation of 1991. Expecting a declining military demand after the end of the Cold War, HumVee manufacturers started producing the $100,000 Hummer for the civilian market in 1992. In the USA the manufacturers failed to sell enough civilian HumVees to make much money, so they sold out to General Motors in 1999. Ford responded by buying Land Rover the following year, already awash with cash from the success of their own in-house SUV programme.

Lutz had launched his Ford Explorer in 1992, driving it through a huge plate glass window at the Detroit auto show that year – and further building the link with violence. The point was not to reflect a growing brutalisation of society, but to target people the market researchers describe as 'especially self-centred'. Thanks to detailed market research (exposed in Keith Bradsher's pivotal 2002 book, *High and Mighty*), and the help of in-house psychologists, the marketers knew a great deal about SUV buyers and how to manipulate them.[15]

Bradsher's book hit a nerve when it was published, because the psychological profiling that the car makers were doing to identify their market wasn't exactly flattering. But, if anything, it certainly chimed with what has become the popular, cultural reputation and stereotype of the SUV owner. The car makers deliberately couched their marketing appeal to people who are more self-centred and vain, who tend to be worried about their marriage and relationships; people who have a weak sense of community but are insecure about how other people see them.[16] Recognising that decisions about car purchases are now taken more by women, they also sought to appeal to people who are more worried about dangerous driving – their own and everyone else's. That was the profile that Bradsher was given by car marketing departments.

Early SUV marketing was psychologically targeted towards specific personality types. Most people hated the brash, aggressive advertising for the enormous Dodge Ram. Only 20 per cent of people liked the adverts, but that minority *loved* them.[17]

With the Hummer, GM began to shift from 'the casual brutalisation of nature deployed in the earlier ads' and towards a 'more sinister articulation of nature and society in which the truck's off-road prowess implicitly figured as a means of protecting oneself against social dangers'.[18] In 2001, GM used Arnold Schwarzenegger to unveil the new H2 in downtown Manhattan on the three-month anniversary of the 9/11 attacks.

Japanese manufacturers have been particularly influential in shaping the marketing messages for SUVs. In 2007, the Japanese car maker Nissan managed to steal a march on their rivals with the launch of the Qashqai, which came out of their European operation. Its immediate popularity convinced them to abandon their old saloon car ranges in Europe in favour of the 'mid-market' SUVs. Over twelve months in 2002, as many as 100,000 SUVs were sold in Germany. Three years later German SUV sales had nearly doubled. Some of these were very large vehicles, categorised as over 2.8 tons. By 2006, the European market for SUVs had reached 7 per cent. Jump forward to early 2023 and they now make up over half of all new car sales in Europe.[19]

Although the SUV craze came relatively late to Europe, all the same narratives and tropes that have been so successful in selling giant cars to Americans have also been used for the big sell here too. Volvo, for example, has been criticised in Sweden for their advertising campaign which used a well-known comedian (Fares Fares) to market their huge XC60 SUV as a vehicle that 'takes care of you'.[20]

But although their automatic system may protect children inside that particular vehicle from accidents, there is no acknowledgement that SUVs are more likely than average cars to kill you in a variety of ways, especially if you happen to be a road user outside the SUV. Their extra weight delivers more blunt force that is more lethal in an accident.[21] Particular design characteristics such as restricted visibility due to height,[22] and the fact that the SUVs' higher position means it is more likely to hit a person's torso rather than their legs[23] also significantly increase risk of fatal accidents. That higher road position has also been shown to increase the risk of vehicles flipping over in crashes, while analysis published in the *Journal of Safety Research* revealed that in the USA, children are eight times more likely to die when struck by a SUV compared to an average passenger car.[24] Then, of course, there is the issue of their typically well above average pollution levels, which raises both acute and chronic dangers more generally – whether you are inside or outside the SUV.

Manufacturers' promise of greater safety is in glaring contradiction to the apparent facts. Even if you, as a concerned parent, fall for the promise that 'inside' the car your child will be safer, sooner or later they will have to join others as a pedestrian or cyclist on the roads that are now riskier due to your SUV.[25] But the *promise*, repeated endlessly in glossy magazines, online, on billboards and between television programmes seems to be an important factor explaining why people have turned to SUVs in such numbers.

* * *

In 1916, Theodor Dreiser wrote *A Hoosier Holiday*, comparing car travel favourably against travel by train, and declaring that it brought the 'prospect of new and varied roads, and of that intimate contact with woodland silences, grassy slopes, sudden and sheer vistas at sharp turns, streams not followed by endless lines of (railway) cars'.[26]

For cultural critics like Dreiser, the automobile was a key tool that helped to mediate between civilisation and nature. Advertising images juxtaposed shiny new cars with scenic vistas. Holidaymakers were shown picnicking in the countryside beside their cars. Garden cities and suburbs blossomed because they were 'within an easy drive' of the big city. In the 1930s, it would take less than 20 minutes to drive from London's West End to the aerodrome at Croydon, a journey that would now take at least three times that, day or night.

This was the genesis of the paradox at the heart of all motor advertising, which has grown more pointed and poignant with every passing year. Although it would not become obvious for some decades, making the dream of the open road central to the sales pitch for car ownership would inexorably turn this dream into a myth. The earliest adopters of this new lifestyle technology, like Mr Toad in *The Wind in the Willows*, would ultimately be the only ones to ever really live the dream of the open road. By the 1950s, the available road space was largely full. Far from providing drivers with an open road and the opportunity to escape into nature depicted in the adverts, punters ended up trapped in a traffic-filled, choking gridlock. But with roads filling up, car makers realised their market was threatened with saturation. How could they keep selling more cars with nowhere to put them?

Growing complaints of 'too many cars' clogging American streets were cleverly flipped on their heads by the distorted mirror of the motor lobby's marketers, turning this now impossible to ignore issue into a new problem – the problem of 'not enough road'. The motorway system and the inner-urban expressways were, in this respect, a series of attempts to get cars out of the cities again. German critics at the time wrote about the paradox of their autobahn system taking people out to nature.[27] A visit by British transport planners to see Hitler's autobahns in 1937 shaped the whole UK approach in the years to come.[28] It is not an exaggeration to say that the Nazis were a key inspiration for the building of the UK's motorway network and by extension, the wholesale motorisation of British society by policymakers. What these (and subsequent generations of) planners failed to grasp was that these extra concrete spaces would simply encourage more people to buy and use cars, and to fill them up in turn. Today, the iron law of 'induced demand' is a well-established transport planning principle; every 1 per cent increase in road network capacity will lead to a 1 per cent increase in traffic volumes, as surely as night follows day. This still hasn't stopped transport authorities from trying to build their way out of congestion though; the 'common sense' perspective of 'not enough road' devised by US car marketers in the 1930s and 1940s still has too powerful a grip on the collective policy psyche.

The advent of the SUV offered car marketers a new opportunity to transcend the growing dissonance between the dismal lived experience of drivers and the magical visions of empty roads depicted in their adverts: the tantalising promise of transcending roads altogether by off-road driving. These new consumer vehicles could be depicted in deserts and forests, on glaciers, beaches and mountaintops, providing a fresh chance to rerun the auto industry's original ideas of individual freedom and escape – and to reclaim them from the great dustbin of advertising past. A key adviser to Chrysler employed to help market the shift to SUVs described a deep American cultural foundation underpinning their appeal; one that seems to feed a fundamental contradiction that allows both the worship of nature and the wild, and behaviour that wrecks it: 'We like to describe where we live as the wilderness, whether it's suburbia or downtown. We have to have wilderness, so that we can point to it, and say, "It is a jungle out there. But I am in the wilderness." If there is no wilderness, there is no America.'[29] Exploiting this tension would prove to be wildly successful.

Ford went to enormous lengths in the 1990s to promote the idea that their new Explorer SUV made families or people who drove them 'bold, adventurous and carefree'.

'The new 4-door Explorer – there are no such thing as city limits', said the adverts.

In October 1999, they launched their No Boundaries campaign, linking the car with rugged outdoor individualists, who were shown hiking, kayaking or rock-climbing, with an Explorer parked nearby.

Two years later, in 2001, Ford chairman Bill Ford took part in a television advert explaining that Henry Ford, his great-grandfather, used to prefigure the SUV by taking his Model T cross-country with various US presidents inside. The rest of the film was a kind of glorification of the outdoor life, as lived by Henry and Bill.

At the same time, Ford dealerships were encouraged to get in some camping equipment. One store even installed some trees and a river running through them. Ford launched a travelling show called the No Boundaries Experience, where children could test-drive mini SUVs.

From May 2002, Ford sponsored a string of outdoor festivals and an all-woman team to climb Mount Everest, then co-produced a No Boundaries TV show which saw contestants hiking up the Arctic Circle.

There is no doubt that this taps into a strain in the American psyche that believes in the outdoor life – and in fact Ford had always tried to promote their cars along these lines, even driving a Model T up Ben Nevis in 1916. But since most Explorer drivers never needed their 4x4 capability, there was also something faintly ridiculous about it.

So is the idea of selling access to the wild in this way, when success would undermine the wild completely. Imagine lines of Ford Explorers queuing to get into the wild. It made no sense. Particularly in Europe, which lacks the abundance of wide open spaces enjoyed by Americans, for whose residents the promise of driving into the wild included an element of pyramid selling – after a few years of everyone doing it, the sale becomes meaningless. So were the Chevrolet Blazer adverts quoting Henry David Thoreau in isolation on Walden Pond. Or the H1 Hummer adverts with the message: 'Sometimes you find yourself in the middle of nowhere. Sometimes in the middle of nowhere you find yourself' (Hummer, 2001).[30] This may be true of course, but not at the wheel of a car. At least not without completely upending the mythopoetic men's movement. Or 'Who the heck needs road maps?' (Nissan Pathfinder, 1998).

This was the essence of the first marketing appeal of SUVs, and $9 billion was spent on advertising between 1990 and 2001 to persuade people that they wanted to escape into the rugged purity of the wild ... Though most of the evidence suggests that they never actually did.

In 2011, the advertising agency Saatchi & Saatchi Sydney won an award for their 'no soft stuff' adverts for Toyota 4WDs. In their own words, this was 'designed to intimidate and interrogate city people, while giving country folk a new champion. "Tofu, hair gel, small fluffy dogs, roller blades, soy decaf lattes. Not on our watch".'

We realise that it was an attempt at populist humour, the kind that kicks down at difference from a position of comfort; but fomenting strife between town and countryside, and between drivers and other road users, feels more dangerous a decade later. It's the kind of populism that also gives tacit permission for aggression of various kinds against the identified targets of derision.

Perhaps because, in the end, marketing messages about loving the wild so much that you asphyxiate it seemed so corny, another message emerged at the same time from Cherokee advertisers trying to drag down Ford's market share. And this message was too dangerous to be corny.

One 1999 advert for the Jeep Cherokee seemed to more than hint at violence. It appeared as an apparent celebration of the destruction of the American West, with pictures of Native Americans and bison and apparently smeared blood. The slogan was 'Adrenalin rush hour', as if somehow General Custer, or similar, had just lost his head when he slaughtered his way to Little Big Horn.[31]

Also hinting at the acceptability of violence was the Daimler-Chrysler advert for the Dodge Durango: 'This baby carries around chunks of your wimpy wannabe [SUVs] in its tailpipe ...' (Dodge, 2000).[32]

Or Isuzu's 'Put the world at the mercy of your whims' (Isuzu, 1998).[33] Isuzu also sponsors international rugby.

Or Jeep's 'Get out there and show mother nature who's boss' (Jeep, 2000s).[34]

Or Captiva's 'Don't let nature make you feel insignificant' (Chevrolet, 2010).

Or the award-winning 'Born confident' campaign for the Volkswagen T-Roc, which sees a brave ram cowed into submission by the car: 'On meeting the T-Roc, he's clearly met his match', said Autocar (VW, 2018).

These are adverts, in effect, for battered people. They take their cues from Ford's idea of the wild, aware that sometimes the environment around us may feel *so* wild that our only options are to fear it – or to dominate it.

It was Honda's channelling of bodybuilder Charles Atlas which revealed that the target of these adverts were the parts of all of us, no doubt, who feel inadequate: 'Now let's see who gets sand kicked in their face at the beach' (Honda C-RV, 2002).[35]

By then rival brands were becoming increasingly menacing:

'Kick derrière' (Lincoln Navigator).[36]

'Jeep Liberty Benefit #12: The power to master all things, on and off the road' (Tracks along the top of a transport truck allow us all to fantasise about the Jeep Liberty's ability to literally drive over things that get in the way) (Jeep, 2002).

'Tread lightly, and carry a big V-8' (Dodge Durango, 1998).[37] (At more than two tons, nobody could accuse Dodge of treading lightly.)

... a 'bold, aggressive stance. (Intimidating, isn't it?)' (Chevrolet Blazer ZR2, from below the radiator, 1997).

'It's like a monster in a horror movie. It keeps coming back meaner and stronger' (Honda CRV, emerging from a swamp, 2001).[38]

Like the Blazer above, campaigns for Hummer and Jeep were increasingly shot from a position just below the front bumper: we are made to feel prostrate before the vehicles. See for example:

'It only looks like this because it's badass.'

'Pretty much every lane is a passing lane.'

These are more than just empty threats; someone is killed on America's roads roughly every 12 minutes, and the death count reached a 16 year high in 2021.[39] That feeling of aggressive safety that is being cultivated in SUV owners will undoubtedly translate into more aggression from other road users. Making yourself feel safer by making the world at large more dangerous is clearly a self-defeating approach that continually ups the ante on our streets.

Keith Bradsher, author of *High and Mighty*, argued that investment in psychology in the marketing of SUVs had led to clear demarcation between different personality types and the appeal of, for example, SUVs versus 'mini-vans':

> A growing body of research by automakers is finding that buyers of these two kinds of vehicles are very different psychologically. Sport utility buyers tend to be more restless, more sybaritic, less social people who are 'self-oriented,' to use the automakers' words, and who have strong conscious or subconscious fears of crime. Minivan buyers tend to be more self-confident and more 'other-oriented' – more involved with family, friends and their communities.[40]

In his book, Bradsher looked at the encouragement of violence through the consumer psychology of Clotaire Rapaille, a French anthropologist who, as mentioned above, has played a consulting role in the design and marketing of SUVs. People's reactions to consumer items, or so Rapaille says, can be divided according to a number of psychological reactions, one of which he calls 'reptilian'. He argues that SUVs are 'the most reptilian vehicles of all because their imposing, even menacing appearance appeals to people's deep-seated desires for 'survival and reproduction'.

'I think we're going back to medieval times'. said Rapaille. 'And you can see that in that we live in ghettos with gates and private armies. SUVs are exactly that, they are armoured cars for the battlefield.'[41]

The implication is that advertising these cars may actually be encouraging this 'reptilian' state of mind.

The UK geographer John Adams became fascinated by risk in motor transport, because the more he studied it, the more he realised how our statistics are too crude to capture it. Ezra Hauer, one of the key authorities in road safety in the world, agreed that the study was in the 'dark ages'. 'There's lots of arm-waving but very little knowledge of what works', he said.[42]

Why should we be three times more likely to be injured in a built-up area, for example, than on an A road? The answer, said Adams, is not that the A road is somehow safer – it is far more dangerous – it has more to do with the *perception* of risk. Vulnerable people get out of the way. That means that safety measures can 'shift injury rates and fatality rates in opposite directions'. That would explain, for example, why it was that although death rates in traffic accidents went down among car drivers once seatbelts became compulsory, death rates among pedestrians and cyclists went up. Presumably because the drivers *felt* safer and compensated by taking more risks.

There are important implications of this for SUVs which make drivers feel safer, primarily because that is what they have been told to feel, reinforced with an elevated driving position which allows them to look down on other road users from the instinctual safety of higher ground. This sensation of safety means that drivers will tend to 'export' their risk, putting other road users at greater risk. It is hard to pinpoint this because the UK does not make figures for accidents involving SUVs easily available. But, as we have seen, recent research in the USA has shown the vastly greater accident and risk of death rate from SUVs to other road users. It means that these SUV adverts directly put other road users at greater risk by successfully promoting that choice of vehicle, and also through their behavioural messaging.

'Most arguments about road safety turn out, on closer inspection, to be arguments about who should defer to whom in situations of potential conflict', wrote Adams in 1987.[43] In those circumstances, the marketing of SUVs has as much effect on those who *don't* buy as it does on those who do. It implies that we are supposed to defer to them. SUVs may make their own drivers safer and more cocooned (until they step out of them to become pedestrians or cyclists), but they will also tend to make other road users feel less safe – as indeed they are.

* * *

'Suzuki likes nature.' Emblazoned on the spare-tyre cover of a Grand Vitara, these words imply that nature is of interest for the owners and manufacturers of four-wheel drives.

So when the Infiniti QX4 suggests we should: 'Leave the city behind. Leave everything behind' (Infiniti, 1999) – this is a gentler message than the call of the wild pioneered by Ford. Yet it is just as misleading.

There are also depths of meaning here, as there were in the New Zealand television commercial for Jeep Grand Cherokee known as 'shake' (Jeep Grand Cherokee, 2001). It showed the driveway of a pretentious suburban home, where the Jeep disgorges its family of passengers, and then wriggles like a dog, splattering the house and the family with thick globs of mud. This is, says the historian William Rollins, a postmodern 'pretence at authenticity'. It shows the car as part of nature and accepted as such, but domesticated.

'My goal is to experience the natural breath of the earth', it says on the back of some Toyota Landcruisers. Or 'A trip with nature is always a special environmental adventure.'

Or on the tyre cover of a Mitsubishi Challenger: 'This car was created to bring you closer with nature.'

Reading these slogans, as Shane Gunster says:

> Suspicion might begin to arise that the writers employed to turn out these covers have only a tenuous command of English. This is confirmed when a Nissan Terrano comes into view sporting a cover which enjoins us as follows: 'Whenever and everywhere, we can meet our best friend – nature. Take a grip of steering'.[44]

These are slogans written for the Australian and New Zealand market by Japanese marketing people.

Did they actually believe them when they wrote them? Or when they commissioned Land Rover's US advertisements with animals in the bush? Or when Toyota covered their back wheel cover with an outline of a whale's tail? Or when the Mitsubishi RVR tyre cover includes a family trio of penguins, above this statement: 'As environmentally conscious people, we are striving to preserve the antarctic region and all of it's [sic] creatures' – and yes, it did include the greengrocer's apostrophe too.

'This cover appears to make unsubstantiated claims about the company's activities', wrote Gunster. 'A cursory search of available sources revealed only two instances of Mitsubishi having anything to do with the Antarctic: one is an allegation of abetting illegal overfishing of the Southern Ocean, and the other is its interest in commercialising rare biological resources of the frozen continent.'[45]

What, for example, did the Land Rover showroom staff feel when they had to dress in safari gear to get the environment message out? What must the environmentalists who had been convinced by this message have felt? Were they as muddled and conflicted as the car companies? As William Rollins says of Ford:

> It churned out hundreds of thousands of Explorers alongside even bigger models such as the Ford Excursion/Lincoln Navigator – a 6,000-pound leviathan that only an advertising executive could claim 'treads lightly' on the earth.[46]

So Ford, in that sense, has been deluding buyers partly by deluding themselves. That excuse may not extend to the creative teams at their advertising agencies, who must know what SUVs are. As Rollins put it, they reflect 'characteristically "postmodern" aspirations and anxieties, and in particular they embody a profoundly contradictory set of desires, an egregiously perverse expression of our postmodern environmental consciousness'.[47]

In December 2000, Land Rover – by now owned by Ford – produced a full-page advert for the Freelander vehicle. It was published in a range of glossy magazines in South Africa, including everything from *House and Garden* to *Men's Health*, *Complete Golfer* and the international edition of *Time* magazine. It was conspicuously absent from a range of magazines aimed at a largely Black readership.

Instantly, it caused uproar. The South African advertising agency TBWA Hunt Lascaris had wanted to show the speed of the Freelander and the way that it can dominate the African landscape.

It chose to do so by having the Freelander race across a 'barren saltpan' between sky and earth, passed a standing Himba woman from Namibia.

The agency felt it had to really emphasise its speed and how the world around the Freelander bends to its force. To achieve this effect, the Himba woman, who is naked from the waist up, has had her breasts digitally manipulated. They have been elongated and pulled sideways, absurdly, as if they are being sucked along by the Freelander's slipstream. To add to the scene a cloud of dust kicked-up by the car surrounds her legs. For final effect she is made to give the car an admiring glance as it passes.

Years later, the advert that 'shocked South Africans because of its racism and sexism' was still being debated. It was condemned by the South African Advertising Standards authority as 'a violation of human dignity that perpetuated gender and cultural inequality'. Diane Hubbard, of the Legal Assistance Centre in Windhoek, Namibia, said: 'It smacked of the same kind of exploitation that occurred during colonial times', and asked, 'Would a white woman in a bathing costume have been given the same treatment?' Tour operators complained that the advert would encourage people in 4x4s to drive recklessly over the Himba's remote ancestral grazing and burial grounds. Land Rover offered a terse, defensive apology, withdrew the advert and dropped the advertising agency.

But decades later advertising for Land Rovers appears as dismissive of the world, and others in it as ever. A 2021 advert for the large, and highly polluting Land Rover Defender reads, 'Life is so much better without restrictions.' It is pictured off-road in a sun-dappled forest. The reference to restrictions is poignant, and contentious, playing on an 'anti-lockdown' sentiment, as the advert appeared during the Covid-19 pandemic, when regulatory measures which restricted personal movement to protect public health were widespread. The extent to which the public were prepared to undertake personal sacrifices for the common good had surprised many commentators, and of course posed an implicit challenge to the ideology of individualism which underpins consumerism in general, and the imaginary of the private motor lobby in particular. The libertarianism which right-wing columnists prefer to characterise as innate human nature, and achieves perhaps its purest expression in the privately owned SUV, had largely given way to a wave of collectivism and mutual aid in response to the crisis. It was imperative to motordom that society should get 'back to normal' as quickly as possible, before things like the appeal of quiet, safe streets with clean air and birdsong could really take root in the public psyche.

The immediate implication of Land Rover's lockdown advert is that their vehicle should be allowed to drive wherever it wants, through protected woodland, across areas of outstanding natural beauty and sites of special scientific interest, or anybody else's back garden, perhaps. But it also activates a far more fundamental frame around the primacy of individual liberty in service of Land Rover Defender sales. Given the car's outsized carbon emissions (up to three and a half times the EU recommended average, depending on the exact model), the 'no restrictions' frame could equally apply to the amount of pollution it causes; or to its enormous, kerb-hogging size (with its back-mounted spare wheel it is too big for a UK standard parking place).

Specifically, Land Rover is promising the ironic right to roam over natural settings and woodland, never mind that it might be an environmentally sensitive area. But the irony remains of course, most will never even see a woodland except in the distance from the motorway. Because the great majority of SUVs – even the biggest, boasting about their off-road abilities – are indeed registered to urban addresses and are more likely to be seen clogging town and city streets. The advert copy reads: 'Understandably, there are still restrictions as life slowly gets back to normal. Not so with Defender, the 4WD vehicle with a capacity to go almost anywhere and do almost anything.'[48] The 'almost anything' includes pollute, trample and take up more than its fair share of space.

In 2018, the advertising agency BBH London won an award for their 'clown' campaign, persuading viewers that their Audi car could help them navigate the peculiar hazards of other drivers. One Los Angeles Hummer driver was asked why he bought it. He replied: 'I call this my urban escape vehicle', he answered. 'Fires, earthquakes, riots. I'm ready.'[49]

But when it comes to marketing to women, they are not being urged to bring out their inner Trump in quite the same way. In practice, that means that the advertising messages tend to lose their macho edge altogether. Instead, the motor advertisers simply concentrate on the risks out there – on the grounds that you can add an extra cocoon by driving a big car. In fact, the opposite is the case.

Toyota was the pioneer of this idea, when they used New Zealand mountaineer Sir Edmund Hillary's quotation to sell cars – 'Everest can be a ferocious place' (Toyota).[50]

But this was turbocharged by a series of ads for the Lexus 470: 'Let nature worry about you for a change', involving an unlikely pack of crocodiles escaping (Lexus, 1998). There were others with charging bulls, white sharks, cougars, etc.[51]

We've already seen how dangerous SUVs are for other road users, but if it seems obviously true that you and your children are safer in a SUV, unfortunately this does not in fact appear to be the case. This is partly because of their weight and momentum and partly because of their higher centre of gravity, which means that they are too big for crash barriers, which seem likely to flip them over rather than stop them.

We have also known for some time what happens when SUVs hit other cars, mainly because of the efforts of *New York Times* reporter Keith Bradsher, who wrote *High and Mighty* after he was asked that question by the *New York Times* Detroit correspondent Glenn Kramon in 1997.

Bradsher was able to use US accident records to demonstrate that SUV drivers certainly don't make other drivers safer (see above). The fact that sharing the road with ever growing numbers of SUVs demonstrably increases the danger of death and serious injury for all other road users may go a long way to explaining how SUVs came to appeal to a much broader consumer constituency than the niche originally attracted to the faux-military glamour of the early SUVs. Thanks to SUVs, an arms race is underway on our streets; one in which only the biggest, baddest cars will survive.

But Bradsher also found, less predictably, that people are even less safe *inside* SUVs. He soon stirred up a storm in the motoring world. 'The criticisms of SUVs infuriated auto executives, who denounced me in speeches and in interviews with other reporters', wrote Bradsher later.[52]

Bradsher found that the occupant death rate was 6 per cent higher in SUVs than conventional cars, and 8 per cent in the biggest ones.[53] Studies in Arkansas and Utah suggest that rollovers are rare, even in SUVs, but they still account for more cases of paralysis than all other causes combined.[54]

These figures suggest that SUVs were probably killing around an extra 3,000 people in the USA a year at that time – more than died at 9/11. Roughly a third of those died in SUV rollovers, and another third from being hit by one. The final third were being killed by respiratory problems because of the extra pollution caused by SUVs. If similar patterns held in the UK, then it could translate to 500–700 extra deaths a year here. It is true that European SUVs today are typically smaller and more aerodynamic than American ones, but the USA may just be further along the same trend curve to ever bigger cars. Heavier cars will always have a longer stopping distance, and while 4WD makes accelerating slightly safer, it has no effect in poor driving conditions.[55]

And if you survive the rollovers, watch out for the air quality. Ford have now produced 7 million Explorers, but – since the 2011 version – they have received over 2,700 complaints about carbon monoxide seeping into the car, including some from police patrols.[56] The complaints include 80 injuries and eleven crashes after drivers lost consciousness. In short, SUVs are no safer for their occupants, and by some measures may be a good deal more risky than other cars.

In 1998, the Sierra Club kicked off a wave of anti-SUV sentiment with a contest to rename Ford's mammoth Excursion; 'Ford Valdez – Have you driven a tanker lately?' was the winning slogan, referencing the infamous huge oil spillage from the Exxon Valdez supertanker into Prince William Sound, off Alaska, and driving home the blatant discrepancy between ads for SUVs and their real ecological impact.[57]

'What car would Jesus drive?' – a television ad campaign in the Midwest – followed in 2002. The Reverend Jim Ball, of the Washington-based Evangelical Environmental Network, said: 'Most folks don't think of transportation as a moral issue, but we're called to care for kids and for the poor, and filling their lungs with pollution is the opposite of caring for them.'[58]

The campaign was greeted with rage of a kind that seems pretty inexplicable to a European audience. But then in Europe, we have a system of permit-trading that has its own kind of perversity, and seems to encourage the kind of doublethink that the American SUV manufacturers have used so successfully to enrich themselves. European regulations specify an average emissions standard across the whole of a manufacturer's production fleet – but they can also buy emissions credits from rival manufacturers to improve this average.

And so it is that the electric car-only manufacturer Tesla generates so-called 'super credits' for every electric vehicle it sells, and then sells these credits on to makers of far more polluting cars to enable them to comply without changing the emissions profile of their own production fleet. For instance, SUV manufacturer Fiat Chrysler, which has to spend up to £500 million a year just to buy super credits from the likes of Tesla. They are doing so because the SUV market remains easily profitable enough to underwrite these extra costs.

This means that, generally unbeknownst to them, buyers of Tesla cars are inadvertently enabling the production of gas-guzzling monsters. Toyota, Citroen and Nissan all offset against the sales of the Nissan Leaf, for example, which is simply putting off the 'evil' day. As one columnist from *Autocar* put it, the only way forward is to make their SUVs more fuel efficient.[59]

Largely because of this elaborate regulatory shell game, the average emissions/km of new cars had, in the years prior to the pandemic, begun rising in both Europe and the UK, after decades of annual reductions.

The other way manufacturers have been trying to avoid the letter of the regulations is by reducing their fleet emissions by inserting small electric motors to produce what are known as 'mild hybrids'. This involves a small 48 volt lithium battery which allows for up to 12 per cent better fuel economy by starting the petrol engine in a more fuel-efficient way. But these are not true hybrids – and fleet emissions are still rising because the petrol SUV market still dominates.

The sheer size of the SUV cars also creates a different problem – big wheels have been turned into aspirational, luxury items. But big wheels mean big tyres, which are more polluting than smaller tyres, because small pieces of rubber and synthetic materials – micro plastics – shed into the environment at a greater rate. Microplastic pollution from car tyres is one of the great, untold pollution scandals of our time. A report for the European Commission concluded that 500,000 tonnes of microplastic pollution were generated from tyre wear every year in the European Union,[60] the annual figure given for the UK was 68,000 tonnes.[61] Stung by these revelations, the tyre industry responded in a manner reminiscent of the tobacco industry seeking to minimise the importance of the threat.[62]

Electrifying cars solves for tailpipe emissions, but may even worsen particulate emissions from tyre and road wear. Part of the problem is that the limitations of battery propulsion technology favour larger, heavier cars. Volkswagen took their small electric e-Up cars off the UK market after only a few months and only 400 sales. Unfortunately for the air we breathe, the Audi e-tron is 2.6 tons – when the average car weighs only half that.

* * *

The cultural historian William Rollins summed up the green paradox of SUV marketing like this:

> What makes the SUV so outrageous is that it is a twisted expression of that developing environmental consciousness, a perversion of energies that might, collectively, have built something far more sustainable by now.[63]

That is a condemnation of the motor manufacturers who created this monster, but perhaps even more so of the in-house marketing departments and advertising agencies which have applied their creative skills to allow this monster to thrive.

Of course, nobody should expect advertisers to tell the whole truth about the products they sell. After all, advertising frequently works precisely to remove the product from its often grim process of production (think factory-farmed animals, shoe, clothing and iphone factories) to create an entirely new, imaginary world of promise and satisfaction for the product, and you, the buyer, to inhabit. We have already looked at the paradox inherent to the way that cars are sold to the public – pretending that people will drive them along wide, empty roads with the wind in their hair, when the reality will mainly be spent in stop-go traffic jams. But the way that creatives have stretched the distance between their storytelling and the objective reality of SUVs elevates this sophistry to a dizzying new heights.

The automobile industry is the heaviest advertiser in the United States by a large margin; it spent close to a billion dollars advertising SUVs each year through the 1990s, to establish the untruths discussed here in the public mind. There is no sign of this abating. The Australian media in 2017 was reporting that – as SUV sales began to overtake passenger cars – the advertising spending for SUVs by manufacturers had risen by 86 per cent in only one year.[64] But the increasing market share of SUVs is now pushing climate goals out of reach. Anyone reaching for the electric 'get out of jail free card' would do well to consider that because of the extra weight and energy intensity of making an electric SUV, this has the potential to actually make matters worse.[65] Electrifying road vehicles is a central and essential aspect of our societal response to the climate crisis. But so is reversing the trend towards ever bigger, more resource-hungry private cars; indeed a 2023 study from the UK Health Security Agency estimates that the potential emissions savings from 'downsizing' new car sales away from large SUVs are of a 'similar magnitude' to those available from electrification.[66] The electric SUV is ultimately as much part of the problem as it is part of the solution.

The explanation of how SUVs have rapidly come to dominate both car sales and city streets set out in this chapter leads to a fairly obvious remedy: that advertising should be ended for SUVs. The tiny share of car buyers who actually need such large and powerful vehicles would still be free to purchase them. But the vast bulk of drivers who do not might be better able to make more sensible choices about their actual mobility needs without the cultural pressure to up-size which advertising applies.

When we launched the Badvertising campaign in 2020, we called for adverts for new cars with emissions exceeding 160g of CO_2 per kilometre, or with an overall length exceeding 4.8m – that's longer than your average crocodile – to no longer be permitted in the UK in any form. These thresholds would equate to an advertising ban on the dirtiest third of the UK car market in terms of carbon emissions – and on all cars which are too big to fit in a standard UK parking space.

Since then, with the announcement of an official pledge to end sales of petrol and diesel cars in the UK by 2030, there is a more direct logic to ending advertising of all such cars. Sometimes there is even a compelling business case to end the promotion of cars altogether: when those adverts are hosted by competing forms of transport, such as rail and metro stations. Organisations like Network Rail and Transport for London are undermining their own businesses by pushing cars to their customers; car ownership is correlated with massively reduced public transport use. Both organisations have clear advertising codes which already prohibit the display of adverts which are against their commercial interests – and therefore they already have a duty to end the advertising of cars on railway and metro stations, bus stops and buses.[67] This should make it easy for public transport authorities in the UK to follow the example of Amsterdam, where the public transport system has banned high-carbon advertising.

More broadly, self-regulation has failed to work so far – the advertising industry's *AdNetZero* initiative is beyond weak. As we'll see in more detail later, the regulation of advertising in the UK (and most other countries) is outdated, insufficiently independent, limited, toothless and under-resourced. But its code does contain one important principle. The code for advertising says that 'Advertising must not encourage behaviour grossly prejudicial to the protection of the environment.'[68] On the basis of available evidence of the harm they cause, the simple act of encouraging people to buy and drive high pollution SUVs, even before we consider the effect of some of the aggressive messaging we have explored here, appears to contravene this principle.

Advertisers have unquestionably facilitated a massive growth in the sales of a technology form that is currently causing thousands of premature deaths and disproportionately fuelling climate breakdown.

We believe that the advertising industry needs to get itself ready for an honest conversation and realistically check everything they do against this principle in an age of climate change. It is an urgent imperative that they stop actively promoting behaviours, products and services that cause harm. We believe there should be a new credo for the creative industries which both firms and individual workers should be allowed to sign, allowing them to reject working on briefs that contravene this principle, without any disciplinary action against them.

6
How Airlines Took Us for a Ride

> Thank God men cannot fly, and lay waste the sky as well as the earth.
>
> —Henry David Thoreau

> Fly green!
>
> —Slogan by Emirates Airlines during the Euros semi-final

Flight has held a special place in the human psyche since at least the dawn of recorded history, and most cultures have myths which revolve around this god-like power. Humanity's attainment of this power through technology in the twentieth century can today be understood as representing the apex of industrial capitalism's

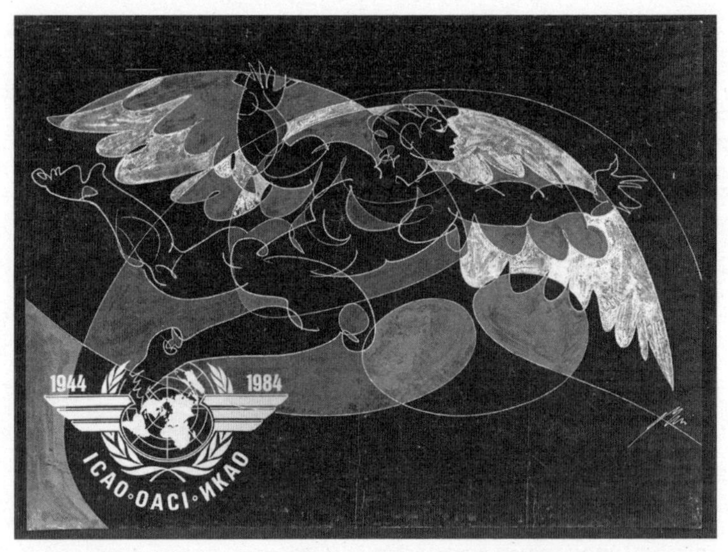

Figure 6.1 The figure of Icarus flying towards the sun was used for the International Civil Aviation Organization's 40th anniversary poster, and later was regularly used by ICAO as cover art for its documents. (Poster by Hans Erni. Source: Smithsonian)

hubris – the belief that humankind has, through its ingenuity, permanently transcended the constraints of the natural world.

This is rather elegantly illustrated by the 'Man in Flight' mural commissioned for the lobby of the first headquarters of the International Civil Aviation Organization (ICAO), depicting the Greek myth of Icarus. The Icarus figure motif from the mural went on to become a regular emblem on ICAO's promotional materials in the 1980s. But ICAO's Icarus imagery depicts him only in the glory phase of his mythic story arc – before he flies too close to the sun, melts his wings and falls to his death. The Icarus myth is famously an allegory warning of mankind's arrogance in pretending to the powers of the gods, and it holds a certain prescience for the status of air travel in the twenty-first century's escalating climate crisis. Icarus didn't know when to stop, like the aviation industry looks towards its own unbounded growth. An overlooked footnote lost in the popular retelling of the Icarus myth is that he wasn't alone, he flew with his father. And, Daedalus, a master craftsman, understanding the tolerances of the natural materials of wood, wax, string and feather warned his son about not getting carried away. It was ignoring this advice and getting carried away that led to Icarus' demise. Daedalus took a lower, more modest flight path and survived, much as the aviation industry needs to radically reduce itself if collectively we are to survive the twenty-first century.

As ICAO have it, 'the history of flight is the history of a dream'.[1] This is potent raw material for modern marketers to work with, and if cars have become consumerism's ultimate positional good (the kind of possession that gives you a sense of your position, or status, among others), air travel has become its ultimate aspirational service. In the context of this book, civil aviation also has a special status for several other important reasons.

First, from the perspective of individual consumption behaviour, flying is the fastest way to fry the planet.[2] It might be possible to have a greater warming impact in a shorter time by starting forest fires or smashing up old fridges, but these are not behaviours which are considered a normal, everyday part of modern life in rich nations. Consequently, of all the possible ways to cut your personal carbon footprint, if you are one of the lucky few that fly regularly today, cutting out just one long haul return flight per year will have the biggest impact.[3]

Second, from a technological perspective, aviation is a notoriously 'hard to treat' sector. Since the turn of the century air travel has been the fastest growing source of transport carbon emissions, and modest fuel efficiency gains have been outweighed by roughly four to one for decades by rapacious growth in demand for flights.[4] By 2019, there were 4.5 billion passenger flights on scheduled services – three times as many as there were in 1997, when the world's nations first came together at Kyoto to pledge to tackle climate change.[5] The ICAO projects that by 2040 there will be 10 billion passenger flights. But so far big promises of green planes have come to little; of over 50 climate targets set by aviation industry actors since the year 2000, just one was actually met.[6]

What is becoming clear is that there are simply no techno-fixes available today, or on the horizon, to enable zero carbon flight on a scale and timeframe that is meaningful for addressing the climate crisis. Modern jet engines are a very mature industrial technology, with further incremental efficiency improvements offering ever diminishing returns. Efficiency gains are, in any case, strictly bounded by the laws of thermodynamics. You can only improve so far before physics gets in your face. There is nowhere left to go but a step change to an entirely new form of propulsion. But known battery technology is too heavy for all but the smallest craft on the shortest routes to go electric; its 'gravimetric energy density' is much too low. Hydrogen is much lighter, but lacks the 'volumetric energy density' required for flight; entirely new aircraft designs would be needed in which most of the internal space is taken up by hydrogen fuel.

In either case, new aircraft take decades to develop and certify for passenger flight, and the slow rate of fleet turnover means it will be further decades after they first enter the market before they make up more than a tiny share of all planes in operation.[7] For context, Boeing and Airbus, the global aerospace manufacturing duopoly responsible for almost every passenger jet in service, between them sold over a thousand new, conventionally propelled commercial jets in 2022.[8] Their long operational lives mean these planes will still be in the skies in 2050, when the world is supposed to have achieved a net zero global economy.

Hence, the industry's hopes are focused on what they call 'Sustainable Aviation Fuels' (SAFs) – non-fossil-derived hydrocarbon substitutes for kerosene which can function as 'drop-in' fuels for standard aircraft engines.[9] But as the UK's scientific authority the Royal Society put it in a recent briefing paper, 'there is no clear or single net zero alternative to jet fuel'.[10] Biofuels from agricultural feedstocks compete with food crops and rarely result in real world emissions reductions, while fuels derived from waste oils cannot substitute for more than around 2 per cent of kerosene consumption. Overall, existing alternatives to kerosene can be scalable or sustainable, but not both.[11]

The most promising long-term option, and one that is relied on heavily in aviation decarbonisation 'roadmaps' to 2050, seems to be e-fuels – synthetic kerosene created from renewable electricity and carbon captured directly from the air. But a recent study finds that even this low-regrets possibility isn't without problems, as the huge amounts of energy input needed and low efficiency of the multi-stage production process could see use of these fuels at scale consume 9 per cent of all global renewable electricity supplies in 2050.[12] But modelling shows that almost every other industrial use of a unit of renewable electricity yields a bigger reduction in carbon.[13] These system-level effects lead the authors to conclude that rather than contributing to efforts to tackle climate change, 'SAF production undermines global goals of limiting warming to 1.5°C.'

Back in the real world, e-fuels remain commercially hypothetical at the time of writing, and even if production can be rapidly scaled over coming years, e-fuel is expected to cost three to five times as much as the kerosene it replaces.[14] Today, in total SAFs account for less than 1 per cent of all jet fuel consumption, and the industry has never met a single one of its previous targets for SAF takeup of any kind. We should also be careful what we wish for; airlines are now demanding vast public subsidies to cover the additional costs of increasing the share of these rarified fuels in their supply mix.[15]

This technical tangent matters to the thesis of this book because of the unavoidable implication that the lack of technological solutions to the warming impact of air travel presents: that meeting climate targets requires us, collectively, to fly less.

This is indeed the conclusion of every major deep decarbonisation pathway produced by anyone other than the aviation industry itself.[16] The UK's statutory advisers on achieving net zero, the Climate Change Committee, have told the UK government to 'Implement a policy to manage aviation demand as soon as possible so the mechanisms are in place in the likely event that low emission technologies are not commercially available to meet the Government's aviation pathway.'[17]

Even the global net zero pathway produced by the techno-centric intergovernmental establishment body, the International Energy Agency, projects that policies to reduce demand for flights will contribute more to reducing emissions from air travel than technological advances in the aerospace sector by 2050.[18] In other words, behaviour change will be more effective than widgets. The UK FIRES academic research programme led by Cambridge University goes further, concluding that 'There are no options for zero-emissions flight in the time available for action, so the industry faces a rapid contraction.'[19]

This is why flying is such a special case in the grand scheme of our efforts to decarbonise. If reducing demand for flights is the only surefire option available to reduce the environmental impact of flying, then it is inescapable that industry efforts to persuade people to fly *more* cannot continue to go unaddressed.

There is a third dimension to aviation's unique status in the climate crisis problematic: the stark fact that it is overwhelmingly a discretionary leisure activity undertaken by a very small share of the global population who are at, or near, the top of the income spectrum. Unlike say, meat and dairy consumption, or indeed car use, air travel is not a daily fact of life for most of us even in richer nations (globally, fewer than one in six people own cars, although over four in five report meat in their diets). Almost all of the environmental damage from plane travel comes from a relatively tiny number of people. In the UK, just 15 per cent of the population take around 70 per cent of all flights, and half of us take no flights at all in any given year. This is a pattern that is repeated in every major aviation market in the world.[20] Flying behaviour is so concentrated that 1 per cent of the global population is estimated to be responsible for 50 per cent of all aviation emissions.[21]

What's more, although many people's mental image of a frequent flyer is likely to be an executive flying on corporate business, in reality, by 2019 business travel had declined so much that fewer than one in ten international flights by UK residents was for business.[22] By contrast, two thirds were simply for holidays, and a quarter for visiting friends or relatives. British residents fly more internationally than the people of any other nation, and in certain social echelons, air travel has become as normalised as eating out in restaurants. Yet studies of frequent flyers show that they themselves regard around half of the leisure flights they take as 'unimportant'.[23]

This extreme energy profligacy by a global elite can be largely explained by the uniquely generous tax treatment of air travel; kerosene is the only form of fossil fuel that remains entirely untaxed by international treaty, and is also, in most countries, 0 per cent rated for the sales tax, VAT, alongside essential items like wheelchairs and baby clothes. All of this makes flying artificially cheap relative to both other forms of transport and other leisure activities, skewing consumption behaviours towards air travel, and meaning its direct costs make up a trivial part of the spending profile of rich households. Remedying this perverse set of fiscal arrangements is outside the scope of this book; the need for an equitable tax framework for air travel that can keep overall flights within safe limits for the climate, while still protecting access to some quantum of air travel for all, is what led us to develop the concept of a frequent flyer levy. But at this point in our story we want to revisit those key facts about aviation as they relate to the role of advertising in driving the climate crisis.

Flying is the most carbon-intensive consumption behaviour available. The only way to bring overall emissions from air travel in line with climate goals is to reduce demand for flights. Demand for flights is overwhelmingly for leisure travel by a small number of wealthy frequent flyers, much of which is seen by the travellers themselves as unimportant. As we saw in Chapter 3, advertising a product increases demand for it and therefore carries the climate and ecological impact of its consumption.[24] Together, these insights suggest that advertising could be an excellent place to look for ways to cut carbon emissions from air travel fast.

So in 2022 we worked with Greenpeace to do just that, on a research project attempting to quantify the estimated carbon emissions generated by advertising for some of the most damaging sectors – namely, cars and airlines.[25]

Based on publicly available data on greenhouse gas emissions, advertising spend and a few selected typical ratios for the returns on advertising spend, we found that globally, car and airline advertising in 2019 could have been responsible for 202–606 million tonnes of carbon dioxide equivalent (MtCO2e). Airline advertising alone was associated with as much as 34 million tonnes of carbon dioxide equivalent – around the size of Wales' annual emissions.

The findings were only a snapshot in time of marketing activities for two exceptionally polluting sectors, unable to fully assess advertising's global climate impact. The research couldn't touch on the degree to which advertising had helped to create the overall market for cars and flights by shaping society; for example, by normalising private car ownership as opposed to public transport and active travel, or flying as opposed to taking the train. Nor could it reflect the impact of historical emissions. Because car makers and airlines do not disclose their actual returns on advertising spending, our researchers had to use a few selected typical ratios for the degree to which advertising drives sales. This lack of transparency around high-carbon companies' returns on advertising spending is a problem, as it shields the advertising industry from greater scrutiny of the environmental impact of their activities. The overall impact of advertising fossil-fuelled lifestyles, products and services is obviously far greater than that assessed in our study.

Nevertheless, the exercise was helpful in substantiating the scale of the link between advertising high-carbon products and increasing greenhouse gas emissions. While the absolute figure for the impact of car advertising is far higher than that for flights, this is hardly surprising when only 5–10 per cent of the global population flies at least once a year,[26] and a typical car may be driven for 14 years before being scrapped,[27] but only needs to be bought once.

But if we consider that the warming impact of a single transatlantic return flight is greater than that of driving an average sized car for a whole year, it becomes clear that ending advertising for flights could in theory deliver very rapid reductions in emissions. Cars are a slow burn purchase, with a large pulse of carbon emitted during manufacture and then a life cycle of a slower rate of emissions over hundreds of thousands of miles before they are scrapped. Buying a car creates a structural imperative to continue to burn fossil fuels for years to come; car drivers do not need to have petrol advertised to them, they will keep on buying it without ever seeing another ad for it.

By contrast, flights are a fast burn item, generating a huge pulse of greenhouse gas emissions, often of many tonnes, over the course of a few short hours – typically to indulge a desire for leisure which could be fulfilled in many different, lower-carbon ways.

Ending advertising for fossil-fuelled cars can therefore be seen as a structural requirement in the move towards a zero carbon economy and society, but doing this today would still leave tens of millions of fossil-fuelled cars on the road for a decade or more to come. Ending advertising for flights today could see a huge and rapid reduction in emissions from air travel, starting tomorrow.

* * *

Growing public disquiet about the environmental damage caused by air travel has not gone unnoticed by the aviation industry. The coining of a new word – Flygskam, meaning 'flight shame' – in Sweden in 2019, against a backdrop of mass global climate protests led by Greta Thunberg, sent a ripple of panic through airline boardrooms, as it was accompanied by a sudden drop in flights, especially domestic flights, at Swedish and German airports.[28] Incidentally, another term began to be heard in Sweden at the same time: Tågskryt, meaning 'train brag', concerning all the people making the conscious choice to travel by train instead of plane.

'Unchallenged, this sentiment will grow and spread', Alexandre de Juniac, head of the International Air Transport Association (IATA), warned around 150 anxious CEOs at their annual conference that year. 'Come on, stop calling us polluters', de Juniac pleaded.

But the enormous technical challenges – and moreover, the enormous financial costs – of actually taking meaningful action to reduce their own emissions has fuelled a marketing drive to persuade customers that they can continue to fly guilt free by simply paying somebody else to cut their emissions instead.

Carbon offsetting, also explored earlier, is the practice of increasing carbon pollution from one source, but paying a fee, usually via a broker, to reduce carbon emissions by the same amount from a different source. Carbon offsetting as an idea has been around for decades, and was first formalised in the United Nations's Kyoto Protocol in 1997, through the Clean Development Mechanism (CDM).[29] It is easy to see the appeal of offsetting for high-carbon industries with limited options to decarbonise in a commercially, technologically or morally viable way, and it has always been positioned as a 'last resort' remedy for the 'residual' emissions left over after industry has done everything else it can to decarbonise activities.

The largely insurmountable barriers to decarbonising commercial flight suggest this is a sector in which offsetting might have an important role to play in the zero carbon transition. But environmentalists have always regarded offsetting with extreme suspicion, fearing that it could be used less as a last resort and more as a 'get out of jail free card' to justify the continuation and expansion of high-carbon economic sectors and their emissions – an *alternative* to within-sector emissions reductions, rather than an addition to them. And this is precisely what has happened with aviation emissions.

Offsetting is strategically counterproductive when considered from the birds-eye, whole-economy view of a civilisation trying to converge on a zero emissions near-term future. That is because climate action has been left so late that there are no longer any 'spare' emissions reductions available – meeting the challenge means doing everything, everywhere, all at once, and paying for a forest to be planted somewhere does not bring carbon-free air travel any closer to becoming a reality. As we'll see below, the scientific credibility of forest carbon offset schemes has gone up in flames, like many of the actual trees planted as offsets.

Offsetting is theoretically unsound in principle too, for the fundamental reason is that it is not possible to prove the counterfactual claim on which any offset is based; that is, what would have happened if the offset hadn't been bought. Pay to protect an old growth forest? Was it about to be cut down? Would it have been protected anyway? Has the same bit of forest protection been sold to someone else already? Pay to plant a new forest? Will it still be standing decades from now, the time needed to sequester the carbon your flight emitted? How can you be sure? The only provably legitimate offset would be an engineered solution that removed a measurable amount of carbon directly from the atmosphere on roughly the same timeframe as the plane exhaust put it there, and indeed this option is now available for purchase from a company called Climeworks – at a price of around $1,000 per ton. This is approximately 100 times the price of an equivalent offset from a low-cost carrier like Ryanair; readers can no doubt draw their own conclusions. Creating a global market in unfalsifiable claims is manifestly absurd, but this is nonetheless the basis of the $2 billion 'Voluntary Carbon Offsetting' market today, which is expected to grow to five times this size by 2030.[30]

Yet we already have decades of real world experience demonstrating that offsetting is also unsound in practice.[31] A 2016 EU study of the UN's flagship offsetting scheme, the CDM, found that 85 per cent of credits sold under the scheme had not delivered their claimed emissions reductions, and only 2 per cent of projects funded had a high likelihood of doing so. In early 2023, a major investigation into nature-based offsets by the *Guardian* newspaper found that 90 per cent of the offsets sold by the world's biggest certifier were 'worthless'.[32] Pointedly, given what we've noted above, these were schemes related to rainforest protection.

None of this has prevented ICAO, the UN body responsible for tackling global aviation emissions that we met at the start of this chapter, from making offsetting effectively its only substantive approach to the climate challenge to date. The Carbon Offsetting and Reduction Scheme for International Aviation (CORSIA) is the central plank of ICAO's net zero efforts.[33] It is a 'global market-based measure' creating a mechanism whereby other sectors can offer to sell carbon credits to airlines to help them achieve 'carbon neutrality'.

We shall return to the proliferation of 'carbon neutral' claims from airlines at the end of this chapter. But first, let us admire the views from the cabin window of a flight through the last hundred years of the aviation industry's advertising journey.

* * *

Even more so than other new technologies offering consumers something unfamiliar, the nascent aviation industry in the 1920s had first to successfully persuade potential passengers that flying was even safe. Aviation pioneers, the Wright brothers, staged their first flight at a place called Kill Devil Hills. Although everyone survived that, the brothers were later credited in 1908 with the distinction of killing the first aircraft passenger, when a military demonstration flight crashed. Little reassurance was given by the suggestion in an early 1913 book, called *Aviation: Its Principles, Its Present and Future* that suggested safety concerns could be dealt with by taking a first aid kit onboard. The same book confidently asserted that planes were very unlikely ever to be used in war for dropping bombs.

So the earliest adverts for aviation had to hammer home the safety message. The experience for brave ticket holders on the early flights was loud, uncomfortable and for many, frightening.

More sophisticated marketing psychology arrived in the 1930s with the advent of air stewardesses, young women in uniforms

Figure 6.2 This 1925 advert by Imperial Airways ('Air travel *is* safe – the principal life assurance offices have recognised the safety of travel by Imperial Airways'). (Photograph: David Boyle)

deliberately modelled on those of nurses to alleviate the public's fear of flying.

> The psychological effect of having a girl on board is enormous. (Comment about the addition of stewardesses from an airline magazine, 1935)[34]

From here, air travel began to grow rapidly amongst wealthier classes in the United States and Europe, and with high ticket prices and a clientele used to the finer things, adverts turned to focus on the comfort and luxury on offer. Culturally, modernism was ascendant and emphasised speed and aerodynamic design, with the former a big selling point of flight, but now airlines were competing for customers on the basis of the passenger experience also.

> Open your paper when you start, and – by the time you have read it – Paris will be in sight. (Imperial, 1938)

Throughout the 1950s and 1960s, square-jawed men in business suits are depicted living their best lives, reclining contentedly or being attended to by smiling hostesses.

Typical slogans of the time made proclamations like 'You're spoiling me!' (KLM, 1958), 'Look how BOAC takes care of you'

Figure 6.3 'A picture of a man in a hurry', sleeping in his airline seat (Capital Airlines, 1961). (Photograph: David Boyle)

(1964), 'What does a man like for supper 20,000 feet up?' (TWA, 1967), and 'Are you sure this is economy class?' (Air India, 1961). Air travel remained so expensive, and its sales pitch so focused on luxury, that this period saw aviation take on an aura of glamour and prestige which persists to this day – at least in the abstract.

The technological miracle of flight now entered a new phase with the arrival of commercial jet airliners, and its unprecedented speed as a mode of transport began shrinking the world for those with access to it. Frequent transatlantic travel became fashionable with the rich, heralding the arrival of a new breed of traveller – the 'Jet Set' – even as larger, faster, more efficient jets saw ticket prices begin to fall, undermining air travel's exclusivity. Passenger numbers more than quadrupled between 1955 and 1972 – but most flights were still being taken by businessmen (the original frequent flyers), so these were the targets of the marketers. This was the period in which the hit TV show *Mad Men* is set, and a newly confident but intensely patriarchal advertising industry turned its attention to transforming the humble air hostess from a reassuring nurse-like figure to a deliberate sex object.

In 1965 Braniff International took on Emilio Pucci to design new uniforms for their hostesses where 'Sex is the message'. Advertising executive Mary Wells, who managed the commission, said: 'When a tired businessman gets on an airplane, we think he ought to be allowed to look at a pretty girl.' The now defunct airline was one of the trailblazers in changing the design aesthetic of the aviation business in the 1960s, producing iconic adverts like their 'Air Strip 1', depicting a flight attendant removing outer layers of garments to reveal a swinging 1960s contemporary outfit beneath. (We asked Braniff for permission to reproduce their Air Strip advert here, but they refused, telling us the ad had been renamed something more innocuous, and that they reject any suggestion they were involved in sexualising air hostesses at the time: 'Braniff more than any carrier enabled our flight attendants to be who they wanted to be and live in the new revolutionary world of the 1960s as they saw fit.')

This triggered an arms race for shorter skirts and recruiting for sex appeal. 'If you like flying American Airlines, but you're not sure why – maybe she's why', a late 1960s advert suggested of stewardess Mary-Anne Dennison.[35] Mary-Anne goes on to explain, 'One passenger, he was very, very nice, he stopped me in the aisle and said he always flew American because they have such pretty girls.' By 1973, South Western Airlines were openly saying that when they interviewed women to be stewardesses, they used to 'start with the legs and work their way to their faces'.[36] 'The girls must be able to wear kinky leather boots and hot pants or they don't get the job', said a company spokesperson.[37]

Consequently, American stewardesses became one of the first groups to file a case under the country's Civil Rights Act, which made discriminatory hiring practices illegal, and over the course of

Figure 6.4 Pacific Southwest Airlines stewardess, 1970s. (Pacific Southwest Airlines)

the 1970s the highly sexualised 'air stewardesses' slowly became the gender-neutral 'flight attendants' who serve air passengers today.

Deregulation of the US airline industry in the late 1970s was hugely disruptive to the incumbent players, but the overall effect was to drastically cut ticket prices, eventually across the entire global aviation market. Freedom to compete on ticket price and growing economies of scale combined with the uniquely generous fuel tax breaks discussed above to create a whole new form of travel – the package holiday – and a whole new type of aviation business – the low-cost carrier. Plummeting air fares were accompanied by the abandonment of luxury and the arrival of narrow seats on crowded planes and congested airports. Some felt the quality of their advertising suffered too. 'Before deregulation, airlines had to apply for rate changes. They had to do much more advertising, focused on things other than price', said Matthias Hühne, the author of *Airline Visual Identity*. 'For example, the destinations in the advertising look much more exciting. And they had to focus on the beauty of the aircraft.'[38]

In the UK, the original idea of the 'package holiday' has its roots in the mid-nineteenth century, when the firm Thomas Cook organised tours for people first within the UK's borders and then to Europe. But in the form we see it today things took off, literally, with changes to the Convention on International Civil Aviation that meant charter planes could be used to provide for mass tourism. For British tourists at least, Spain became a major destination with one region even rebranded as the Costa Blanca for marketing purposes to promote it. The popularity of flying on package holidays to the Mediterranean grew from the 1960s onwards, reshaping Spanish and other coastlines as sleepy fishing villages turned into vast avenues of concrete hotels facing increasingly crowded beaches.

The next step came with the rise of the budget airline, or 'low-cost carrier' in the early 1980s. In the USA they were boosted by the airline deregulation of 1978 and the demise of the Civil Aeronautics Board in 1984 which had closely managed the industry, while protectionism amongst European countries meant deregulation took ten years to complete from the start of the process in 1987.

Airlines like Pacific Southwest, New York Air and Jet America set the model and the relative cost of flying plummeted. In the mid-1990s Europe's first low-cost carrier EasyJet copied and adapted the business model, and was soon followed by an influx of 'no-frills' rivals. Brash and bold marketing focused on price and, because they took landing slots at less well-known regional airports, was often slightly misleading about exactly how close to your advertised city destination you would land. EasyJet targeted their ads at a youthful 'generation easyJet', and focused on normalising flying on a whim:

> The I can't wait to go generation.
> The early risers for the airport cab, last minute packing, full of excitement generation.
> The head first into water, wine or work generation.
> The nip over, seal the deal, back for story time generation.
> The walk until you're lost, find a quaint spot, strangers become friends generation.
> The we've been coming here for years, but still fall in love generation.
> The I don't want to go home, let's stay longer generation.
> The back at the office, staring out the window, let's do it all again generation.
> The everyone doing it their way generation.
> The more places, more choices, more often generation.
>
> This is generation easyJet.
>
> —David Beattie, Generation EasyJet advert, 2013

UK passenger numbers skyrocketed from around 17 million in 1985 to nearly five times as many in 2010. While the demographics of air travel had changed, many of the aviation industry's leading public figures were as unreconstructed as ever. Virgin's Sir Richard Branson loves nothing more than posing for photos with a beautiful flight attendant on each arm, and during the 2000s, poor taste stunts aimed at deliberately courting controversy became a de rigeur part of the aviation marketers' toolbox. When British Airways was running into trouble lifting their London Eye ferris wheel in 1999, Branson hired a balloon to fly over the Thames with the slogan 'BA CAN'T GET IT UP!'

EasyJet's arch-rival Ryanair's long serving chief executive, Michael O'Leary, oversees much of the promotion of the airline himself. Ryanair uses cheap, simple adverts that tell passengers that Ryanair has low fares – and O'Leary uses controversy to promote the business. In 2009, he suggested that passengers would be charged £1 to use the toilets onboard, explaining that passengers could use the terminals at either the destination or arrival airport. This would speed things up, he said. O'Leary also argued that larger passengers should be charged more since they took up more room. The resulting furore won the airline more exposure than a multi-million pound ad campaign would have achieved.

In a similar vein, a newspaper advert for EasyJet, published in the *Evening Standard* during the Iraq war, showed a woman's breasts in a bikini top beside the line 'Discover weapons of mass distraction', attracting hundreds of complaints that it was demeaning to women and trivialised the war. (The Advertising Standards Authority cleared the ad, concluding it was 'light-hearted and humorous' and unlikely to cause offence.)[39]

By the end of the noughties, the sheen was starting to come off mass air travel, as the world began to wake up to the growing threat of climate catastrophe. In 2008 the UK government passed the landmark Climate Change Act, and by the end of 2009 their new statutory advisers, the Climate Change Committee, were warning that because there is such limited scope for technology to decarbonise flight, 'deliberate policies to limit demand below its unconstrained level are therefore essential'.[40]

The global financial crash underway led to a brief flatlining of aviation demand, but flights soon ticked back up, eventually reaching the 4.5 billion passenger peak of pre-pandemic 2019. As discussed at the outset of this chapter, 2019 was also a new high watermark in climate concern. But in 2020 the burgeoning grassroots global climate movement and the rapidly expanding aviation industry eyeing it warily would both be stopped in their tracks by the Covid-19 pandemic.

Flights were grounded at an unprecedented scale in 2020, but not before the aviation industry had ensured the contagion had spread to every inhabited continent on the planet. US airlines successfully lobbied for huge public bailouts which were intended to protect jobs, but frequently wound up lining the pockets of shareholders.[41] Some European airlines went bust, while others fired employees en masse, only to scramble clumsily to rehire them on worse contracts when demand for air travel returned.

In the aftermath of the pandemic, there is now a massive effort on behalf of airlines to push governments and airports around the world to roll back environmental taxes and set up bailout funds after the coronavirus crisis. Even as the world's borders closed in 2020, a strategy document produced by the airline industry shows it lobbying for public money to be poured into funds to restart or maintain air travel, and for any planned tax increases to be delayed. The International Air Transport Association (IATA) document detailed lobbying strategies for Europe, the Americas, Asia and the Pacific, North Asia, Africa and the Middle East.

The IATA lobbying guidance, which was mistakenly uploaded to the backend of the group's website and removed when environmental investigative outfit Unearthed enquired about it, recommends airlines lobby governments and airports for, among other measures: the immediate reduction of all charges and taxes; deferral of any planned increases in charges and taxes for 6–12 months; setting up of funds to help airlines restart or maintain routes; reducing airport staff numbers to save money rather than increasing airline charges.

Regardless of how well this lobbying works, the post-pandemic aviation industry now faces a more existential problem as it seeks to return to its rapid growth trajectory. Marketing airlines began by stressing that high speed, high altitude travel in a new technology was safe, and moved on to promoting the idea of flying as joining the 'jet set', hyping the cosmopolitan glamour and the sex appeal of travelling by air. Much of this aspirational aura around aviation has persisted into the age of mass air travel, even though the actual experience of flying as glamourous remains as elite and unattainable as ever for most. The noughties saw efforts focus on the normalisation of frequent flying, and a turn to meta-marketing tactics to reach cynical younger generations who had grown up inured to less sophisticated appeals. In the 2020s, if the travelling public begin to see air travel as inextricably linked with environmental destruction, there is a very real possibility that they will turn away from flying altogether. With no real answer to the charge that they are a highly polluting industry that is driving climate breakdown, airlines have therefore placed their efforts in promoting a fake one: 'carbon neutrality'.

In December 2021 United Airlines made a triumphal announcement on social media that it had launched the first aircraft powered with 100 per cent sustainable aviation fuel. This news followed United's promotion of its Eco Skies programme, a partnership scheme whereby corporate clients can sign up to help fund the increase in sustainable fuel while lowering their carbon emissions. Adverts appeared in which the airline depicted themselves as 'Sky Huggers', with pictures of a plane in flight seen through trees so that the view can't avoid the association with the environmentalist stereotype of 'tree hugger'.

On closer inspection, the claim made by United's marketing team is highly misleading. Only one engine was powered by 100 per cent 'sustainable' aviation fuel (SAF), while the other was filled with a 50/50 mix of allegedly sustainable fuel and regular jet fuel. This key detail was noticed and was picked apart by Twitter commentators,

Figure 6.5 United Airlines, 2021. (Photograph: Andrew Simms)

who were quick to call the United's announcement a greenwashing stunt.[42]

On the company's official website, no information is to be found on what the company's sustainable aviation fuel is sourced from.[43] Twitter users promptly enquired and United's official account responded that only second-generation biofuels – derived from 'waste' products – were used. But can the airline's own claim in a tweet with no backing by factual sources nor third-party independent verification be trusted? When asked about what companies produced the biofuels, United replied that it had used a majority (80 per cent) of HEFA – hydroprocessed esters and fatty acids – from World Energy and 20 per cent of SAK – synthesised aromatic kerosene – produced by the company Virent.

The 'Skyhugger' ad and its tenuous relationship with the truth is emblematic of the new focus of the aviation industry's advertising strategy in the 2020s. At the football clashes at Wembley Stadium in London during UEFA's Euros 2020 (which due to Covid-19 actually happened in 2021), there were many prominent flashing pitch-side advertisements from high-carbon sponsors. They included ones for Qatar Airways, the international airline that recently became the largest air-freight carrier in the world: 'QATAR AIRWAYS – FLY GREENER'.

We made this spurious message the subject of a complaint to the Advertising Standards Authority (ASA) in the UK, but they sidestepped it, claiming that they had no authority over ads in that context, and referred us instead to Ofcom and UEFA – a common pattern of passing the buck for those of us seeking accountability for misleading green claims by polluting companies. In November 2021, the ASA also refused to investigate ads by airline EasyJet, designed by ad agency VCCP London,[44] that promoted ticket sales using greenwashing claims about 'Zero Emissions Flight' and carbon offsetting schemes.

The ASA told us that because it was undergoing a review of green transport claims it could not investigate ads by specific companies. The complaint in question was passed to another regulator, the Competition and Markets Authority (CMA) in April 2022.[45] After that, things went quiet, despite the CMA claiming to be cracking down on misleading green claims. During 2021–22, the ASA also refused to investigate green claims by major polluters including Chevron, McDonald's, Standard Chartered Bank and Barclays (the latter of which is the biggest provider of fossil fuel finance among UK banks, $190.5 billion between 2016 and 2021[46]), as well as any green claims made in ads by the fast fashion and energy sectors.

However, change is now in the air. When the airline Lufthansa offered customers the promise of 'carbon neutral' flying, in line with the IATA plan, the Swedish Advertising Regulator, Reklamombudsmannen (RO), ruled that the advert was greenwash and that it 'contains misleading claims about carbon neutrality' and, as a result, 'that an average consumer risks being misled about the climate impact of air travel'.[47] In the Netherlands, climate campaigners Fossielvrij NL (Fossil Free NL) are in the process of suing Dutch airline KLM over their 'Fly responsibly' greenwashing ad campaign, with KLM's lawyers making a claim that is familiar from tobacco companies before them that 'KLM is advertising to people who will be flying anyway.'[48] Legislators at the European Union are also finally stirring, with the EU's Green Claims Directive and Empowering Consumers for the Green Transition file looking likely to lead to a ban on businesses claiming to be carbon neutral by virtue of their offsetting schemes, thanks to the inability of the scheme's providers to substantiate their claimed emissions reductions.[49]

In the UK too, watchdogs are finally starting to take the problem of greenwashing seriously. In February 2023 the ASA ruled against another Lufthansa advert that carried the promise, 'Connecting the world. Protecting its future'. The ASA found, however, that 'there were currently no environmental initiatives or commercially viable technologies in the aviation industry which would substantiate the absolute green claim "protecting its future"'.[50] The watchdog quickly followed the finding against Lufthansa with another ruling against green claims made by the airline Etihad. Two of its adverts had boasted about taking a 'bold approach to sustainable aviation'. But the ASA concluded instead that there were:

> … currently no initiatives or commercially viable technologies in operation within the aviation industry which would adequately substantiate an absolute green claim such as 'sustainable aviation'.[51]

The ASA's director of complaints and investigations, Miles Lockwood, explained that: 'Our rules make it clear that any green claims need to be backed up with robust evidence and we won't hesitate to ban ads that are misleading.' If consistently applied, this would mean the term 'sustainable aviation' disappearing from all airline adverts. These two rulings may signal a major change of tack in the watchdog's approach to proliferation of greenwashing claims, as the ASA has just announced it will begin much stricter enforcement on the use of terms such as 'carbon neutral' in corporate advertising.[52]

But the ASA may struggle to fulfil this goal in its current incarnation. One of the problems with the way advertising watchdogs work is that they typically shut the barn door after the advertising horse has bolted, using a reactive rather than a proactive approach to enforcement. They also lack powers to impose penalties, are slow to respond, investigate few of the complaints they receive, and, crucially, lack adequate resources to pursue infractions. They also lack consistency. As we saw above, in July 2021 the ASA refused to investigate adverts by Qatar Airways that made the sole unsubstantiated claim 'Fly Greener' to millions of viewers via the advertising boards at the UEFA Euros 2020.[53] The next chapter looks in much more detail at the problems with advertising's watchdogs and how they might be addressed.

7
Why Self-Regulation Isn't Working

> Regulatory bodies, like the people who comprise them, have a marked life cycle. In youth they are vigorous, aggressive, evangelistic, and even intolerant. Later they mellow, and in old age – after a matter of ten or fifteen years – they become, with some exceptions, either an arm of the industry they are regulating or senile.
>
> —John Kenneth Galbraith, *The Great Crash 1929*

That was Galbraith's verdict on regulators, written in a book about the worst failing of regulation in history – the Wall Street Crash in 1929, which led directly to the Great Depression and, in time, to the Second World War.

This chapter looks at how regulation actually works in advertising with a particular focus on the UK as a case study and, for comparison, a roundup of some other national regimes in Europe and the USA later on. It is about the weakness of regulation of the advertising industry generally, which in the UK is mainly undertaken by an industry-funded and industry-staffed body, the Advertising Standards Authority or ASA. Although sometimes unclear, complementary and confusingly overlapping roles are played by several other bodies. One such is the official government Competition and Markets Authority (CMA), which is supposed to guard against anti-competitive practices and, for example, has a watchdog role concerning the control of hidden advertising on social media platforms (a role also played by the ASA).[1] Others include Trading Standards Services, based in local authorities, whose job it is to enforce consumer protection law, and can escalate action if breaches of advertising codes might also break those consumer laws (cases are only very rarely brought), and Ofcom, the government's Office of Communications, which is responsible for the overall regulatory framework for the communication industries. Confusing in theory and in practice. But, broadly speaking, all are charged with ensuring that paid for advertising is clearly seen as such, that it is truthful and not misleading, and that it complies with advertising codes developed by the industry for itself – known lengthily as the UK Code of Non-broadcast Advertising and Direct & Promotional Marketing, or 'CAP' for non-broadcast media, and the UK Code of Broadcast Advertising (BCAP) for broadcasters.

What do the codes actually say? They cover several, general themes. First, and if genuinely applied advertising would look very different than it does today, they call for compliance with the principle that all ads should be 'legal, decent, honest and truthful', and they should comply with the 'spirit' as much as the letter of the advertising codes.[2] At the heart of this is that advertising should not mislead, 'exaggerate or omit key information'.

Hence, in one rare, even groundbreaking ruling in response to a complaint brought by the Badvertising campaign and Adfree Cities, the bank HSBC was found to be at fault for boasting about its sustainability in adverts, while omitting to mention its substantial financing of polluting fossil fuels.[3]

Adverts are also not supposed to cause 'harm and offence'. This, of course, when it comes to offence, is a shifting target based on social norms. And, again, fully implemented such a principle would mean that the job of the Badvertising campaign was already done. Advertising that promotes high-carbon products and behaviours, and in doing so fuels global heating, undeniably causes harm, but currently is deemed perfectly compatible with the codes. This part of the codes specifically says that ads should not encourage 'anti-social behaviour' – perhaps like driving huge, dangerous, polluting SUVs in built-up areas. As we've seen, language on the environment has recently been enhanced, specifically to say that 'Advertising must not encourage behaviour grossly prejudicial to the protection of the environment.'[4]

Yet, in spite of these theoretical boundaries, checks and balances, we remain surrounded by adverts encouraging behaviour grossly prejudicial to protecting the environment, from general encouragement to keep consuming like there's no tomorrow, to the promotion of flying, eating red meat and car makers peddling SUVs as a lifestyle choice. And even when people do complain, the chances of it changing anything are pretty slim.

Of the 21,290 cases of adverts that were subject to complaint in 2022, 503 were for misleading environmental advertising. Out of these, less than one third, only 28 per cent were even investigated, and just 2 per cent of complaints were actually upheld. But the problems go deeper than the very low rate of successful complaints. The worst that generally happens is that an advertiser receives a public rebuke and is compelled to withdraw the offending advert, and typically long after the advert has been widely seen. More serious consequences like fines are even more rare, and happen under different circumstances if an advert has gone beyond falling foul of the advertising codes and broken consumer protection laws.

When the UK subsidiary of a South Korean car maker promoted its hydrogen-powered Nexo car in 2019, it claimed in adverts that it was 'so beautifully clean' that it 'purifies the air as it goes'.[5] Somehow the car maker forgot about all the other types of toxic air pollution the vehicle produced, such as particulates from tyre and brake wear. The ASA ruled that the ad was misleading, but the only penalty was an instruction that the advert should not be used again. That could be set to change with the CMA taking on a wider brief to challenge greenwashing, with the likely powers also to target guilty companies with substantial financial civil penalties.

But, as things stand, in terms of rebukes and desist orders from the advertising watchdog, for a major multinational it amounts to little more than an inconvenience. A new campaign, often a variation on the original theme may already be being rolled out. Shell in the Netherlands, for example, switching from the language of carbon 'offsets' to carbon 'compensations', would be just such a case.

Similar problems of weak advertising regulation can be seen in places as diverse as the USA, the Netherlands, Belgium, Sweden and France. The UK's ASA is the main focus here, and the aim is to bring the regulation of advertising in line with government commitments on meeting climate targets.

The advertising industry has two things to act on. First, like every other sector, it needs to cut its own pollution in line with climate targets. But, more importantly, it needs to address the much larger impact it has by fuelling the consumption of heavily polluting goods and services, which are major drivers of the climate and ecological emergency.

* * *

> Commercial advertising remains mostly self-regulated. This situation is unsatisfactory, and I call on states to adopt legislation. (Special Rapporteur of the UN Human Rights Council)

Galbraith complained about the process by which regulators tend to become ineffective just when you need them most.

In the case of advertising, this *is* a critical moment. If advertisers devote their creative skills and influence to encourage people to fly or buy huge SUVs without another care in the world, just as the planet is threatened by these and other high-carbon products and services, that is hardly decent nor honest.

It is as bad, if not worse, if they also encourage people to believe the greenwash of big oil or gas companies that claim they are working hard to tackle the problem – when they are working even harder to delay action. Or should they try to manage their image and reputation and put themselves in a good light, by sponsoring sports that are healthy outdoor activities, just as the same fossil fuel companies are wrecking the very climate that sport depends on.

These issues need confronting head on, instead of industry regulators falling back on whether or not the advertisements are technically legal, or whether or not they have jurisdiction.

The stories set out in the next sections explain why we believe the ASA has become too close to the industry that funds them and which they are also supposed to be regulating.

* * *

> The interest of [businessmen] is always in some respects different from, and even opposite to, that of the public ... The proposal of any new law or regulation of commerce which comes from this order ... ought never to be adopted, till after having been long and carefully examined ... with the most suspicious attention. It comes from an order of men ... who have generally an interest to deceive and even oppress the public ... (Adam Smith)

The ASA ought to be an important ally in the struggle against the promotion of high-carbon goods and services. But, despite their protestations, and even allowing for a few cases in which some of the most egregious advertising by airlines has been ruled against, that seems not to be the case, and it is possible to see why. Instead of primarily protecting the consumer and society, they have become protective of the industry that they regulate. A key problem is that high-carbon advertising could be the most lucrative work the industry gets.

Here we look at three adverts for heavily polluting brands for which the ASA received complaints.

THE LAND ROVER STORY

The ASA received as many as 96 complaints about an advertisement we briefly mentioned earlier, in the *Guardian* on 12 June 2021 for a heavily polluting, oversized Jaguar Land Rover with a picture of the SUV enjoying itself in a sun-dappled forest. The headline, which for context appeared during the period of pandemic-related lockdowns, said: 'LIFE IS SO MUCH BETTER WITHOUT RESTRICTIONS'.

The text below the image said:

> Understandably, there are still restrictions as life slowly gets back to normal. Not so with the Defender, the 4WD vehicle with a capacity to go almost anywhere and do almost anything ... a whole new world of freedom awaits.

Figure 7.1 Land Rover advertised its large, heavily polluting Defender during the Covid-19 pandemic, exploiting sentiments about health-related restrictions, and implying that their cars could be driven without restriction in sensitive natural woodland. (Photograph: Andrew Simms)

The Badvertisng team told the ASA that the claim 'Life Is So Much Better Without Restrictions' was misleading, because it implied that this particular vehicle was above restrictions or rules, which could cover where and how fast it is safe to drive the SUV, how dangerous the vehicle might be to other road users and its occupants, or restrictions aimed at preventing air pollution and climate change. Like so many other motor adverts in the past, it showed a car alone on the road in nature – whereas, in reality, as we know, most cars spend most of their time in polluted, urban gridlock.

Initially it appeared that the regulator did consider that the ad encouraged and condoned the use of a vehicle in a way that was detrimental to ecologically sensitive environments and was therefore socially irresponsible. But, ultimately, the ASA's council, the panel that decides whether advertisements have broken the advertising rules – and therefore have to be withdrawn or changed, without giving an explanation – concluded there was no problem with the advert.

THE EASYJET STORY

When EasyJet launched a series of ads promoting the airline's *Future Flying* programme, claiming: 'We are championing a future of zero emissions flights', with the bold headline 'Destination Zero

Figure 7.2 EasyJet's advert made the kind of vague, unsupported environmental promise that regulators belatedly are clamping down on. (Adfree Cities)

Emissions', the term 'zero emissions' was not actually defined. But the advert directed the viewer to a webpage where airline tickets were for sale, and you could read about the company's technological hopes for how things might turn out in the mid-2030s.

This was a bit like a company selling a pack of full tar cigarettes today on the promise that cigarettes in 15 years time will be zero tar.

EasyJet's *Future Flying* webpage is headed with an invitation to book a flight, showing that EasyJet is using its pro-environmental messaging to encourage climate-conscious citizens to travel by plane. The perversity, some would say duplicity, is obvious. They are promoting one of the most polluting forms of transport while allowing customers to justify flying under a false impression of environmental sustainability, choosing to fly with EasyJet over other airlines, and choosing to fly more than they would have done otherwise in absolute terms.

In 2019, the year before the pandemic, EasyJet's flights deposited almost 3.2 million tonnes of carbon dioxide equivalent into the atmosphere.

The campaign group Adfree Cities said that EasyJet has no real ambition to reach 'Destination Zero Emissions' on a timescale of any environmental and climate consequence, and in the meantime, its advertising is accelerating destination climate breakdown. After Adfree Cities complained to the ASA, they were astonished to get a reply that rebuffed their complaint because the watchdog was in the process of developing how it would regulate environmental claims in several priority areas, one of which would be the travel sector. In the meantime, it seemed they were powerless to do the job that they existed for. The only reassurance Adfree Cities received was being told that their complaint would provide useful intelligence for the future.

In the meantime, EasyJet continues to make claims about offsetting emissions[6] that are likely now to fall foul of the European Parliament's proposed bill to prohibit advertising offsetting on the grounds that is fundamentally misleading about environmental harm.

THE HSBC STORY

In the run-up to the COP26 climate summit in Glasgow in late 2021, a series of adverts by the bank HSBC were seen claiming to be tackling climate change in ways that appeared both dubious and to leave out certain key facts.

One of the adverts boldly declared that 'HSBC is aiming to provide $1 trillion in finance and investment to help our clients transition to net zero.' While another promoted HSBC's involvement in a tree planting scheme that 'will lock in 1.25 million tonnes of carbon'. And a third one touted the company's ambition to bring its operations to 'net zero globally by 2030'.

But campaigners at Badvertising and Adfree Cities were not impressed by HSBC's version of events, because several key details were missing, and the advertising codes clearly state that adverts should neither be misleading nor omit relevant information. The

Figure 7.3 HSBC was found guilty of greenwash when challenged, for leaving out key information about their large-scale investments in fossil fuels. (Adfree Cities)

complaint sent to the ASA included important details that HSBC, it seemed, would rather not be so public about. For example, in terms of the climate polluting sectors that it financed, the bank was the 13th worst bank in the world, and second worst in the UK.[7] Its supposed investments into tree planting schemes paled in comparison to its more than $110 billion directed to fossil fuel companies between 2016 and 2020, with $23.5 billion to the fossil fuel industry in 2020 alone.[8] While claiming to reduce its operations to net zero by 2030, the company was funding projects like the Carmichael coal mine in Australia, which is predicted to emit more than 4.6 billion tonnes of carbon over its lifetime.

Following 45 complaints, the ASA decided to investigate two of the adverts complained about. The process took nearly a year for the ASA to finally make a ruling, long after the climate conference had drifted into memory, and long after HSBC had taken the benefit of being able to present itself as a climate-responsible citizen to decision-makers and the media during a critical moment for policy development and action. Finally in October 2022, the ASA found that the adverts were misleading by omission. In its ruling, the ASA specified how they informed HSBC that any future marketing communications featuring environmental claims should be 'adequately qualified' and 'not omit material information about its contribution to carbon dioxide and greenhouse gas emissions'.

But again, while this lengthy process was going on, it seemed that the ASA was incapable of investigating other, similar breaches of its codes. In July 2022, when the HSBC ruling was still pending, the watchdog refused to investigate similar, well-evidenced complaints over advertisements by two other major banks who also finance fossil fuel companies, Barclays and Standard Chartered Bank. The reason given for not investigating these was that the ASA were 'currently formally investigating a different advertiser [HSBC] for similar issues'. Which is a bit like the police saying, 'sorry we can't investigate who burned down you and your neighbours houses because we're looking into another arson further down the street'.

And, although the ruling against HSBC was significant and welcome, and led to the bank making public statements to discontinue fossil fuel financing (later reneged on), there are reasons to doubt that it will be sufficient to prevent future corporate greenwashing from big polluters and their financiers.

HOW ADVERTISING GETS REGULATED IN OTHER COUNTRIES

Similar problems to the UK's arise across Europe and in the United States. First, the self-regulatory model adopted by most authorities acts as a significant barrier to implementing new checks and balances fit for dealing with the climate emergency and the greenwash swirling around it. All six countries we look at in Table 7.1 prioritise a system of self-regulation, whereby the ad industry in effect gets to write its own rules. In the USA, France, Sweden and the UK state-funded parallel authorities responsible for consumer protection complement the work of the advertising watchdogs, but their enforcement mechanisms are fairly limited.

At the European level another self-regulatory body, the European Advertising Standards Alliance (EASA), administers the code drawn up by the constituent parts of the industry. Once a complaint against an ad has been filed by the public, a group or company, a jury drawn from various media representatives is then responsible for interpreting the codes.

In Sweden, some of the codes put in place by its watchdog, Reklamombudsmannen, have great potential to tackle harmful advertising. However, Swedish campaigners flag that these are not applied as they should be. The watchdog has a preference for a non-confrontational approach with the industry which harks back to the tendency of regulators to become tame and part of the industry they are supposed to be regulating. This is a general trend across the countries listed in the table of international comparisons (Table 7.1).

The other systemic problem is that these regulators typically lack meaningful enforcement powers. A little, mild public shaming is rarely sufficient deterrent to offending advertisers to continue misleading the public. An oil company like Shell being a prime

Table 7.1 A country-based comparison of advertising regulatory systems

Country and full name of regulator	*Type of regulation and funding*
Advertising Standards Authority (ASA) (UK)	Self-regulation. Funded by the advertising industry.
Consumer and Markets Authority (CMA) (UK)	UK government body
Autorité de Régulation Professionnelle de la Publicité (ARPP) (France)	Industry-funded, self-regulation
Direction Générale de la concurrence, de la consommation et de la répression des fraudes (France)	French government body (Ministry of Economy).
Federal Trade Commission (FTC) (US)	Federal government, state-funded.

Good points	*Bad points*
Wishes to improve its role regarding climate breakdown.	Funded and staffed by advertising industry personnel. Not independent. Ideologically committed to consumerism model. Won't investigate if a complaint falls outside its codes and sometimes narrow remit.
Produced a Green Claims code to tackle greenwash advertising. Has enforcement powers.	Under-resourced to address the scale of the issue.
Because of the organisation's origins in 2008, it is explicitly environmental. No mandate to administer fines. Reviews adverts based on filed complaints (a posteriori) except for TV adverts.	The ARPP is run by representatives of corporations, advertising agencies and the media. Out of the 32 seats, only two are taken by members outside the private sector. Civil society isn't represented. ARPP rulings do not have binding power. They tend to be politically influenced and often lack coherence.
Has enforcement powers.	Has hardly used its power to sanction greenwashing so far.
Has broad authority to look at practices and issue fines. Produces the 'Green Guides', an advisory set of guidelines for avoiding greenwashing. Has the authority to turn these into formal rules, but can't make enforcement action based on the guides. Focused on consumer harm and the impact of claims on spending decisions, rather than simply adjudicating truth/misleading claims per se.	Recent Supreme Court decision has removed the agency's ability to seek monetary damages against advertisers (damages will typically be significantly greater than fines). Can be conservative in defining 'commercial speech', which must be shown to impact consumer/business decisions. Reluctant to go after advertising that may have a political/social aspect. Slow process, with commissioners appointed by presidents and confirmed by Senate. Often led by partisan political agenda.

Country and full name of regulator	*Type of regulation and funding*
National Advertising Division of Better Business Bureau (US)	Industry-funded, self-regulatory.
Swedish Consumer Agency (Sweden)	State-funded and controlled by the government.
Reklamombudsmannen (Sweden)	Self-regulation, funded by industry.
Jury voor Ethische Praktijken inzake Reclame/Jury d'Ethique Publicitaire (JEP) (Belgium)	Self-regulation, funded by industry.
European Advertising Standards Alliance (EASA) (EU)	Self-regulation, funded by industry.
Stichting Reclame Code (Dutch Ethical Board for Adverts) (Netherlands)	Industry-funded, self-regulation.

Good points	*Bad points*
Moves more quickly than FTC. Implements FTC regulations like the 'Green Guides'. Increasingly focused on greenwashing issues and on hearing from non-profit organisations.	Decisions are non-binding (if advertiser refuses to comply with decision, matter can be referred to FTC). Filing fees for having complaints heard can be steep. Tendency to issue 'split-the-baby' decisions to avoid alienating industry players.
Has power if they wish to use it, some good writing in legislation. Publishes guidelines etc. on 'green' advertising.	Hesitant to use the power it has, prone to 'dialogue' and compromise. Slow.
Advertisers have to be able to prove their claims. Quick.	No mandate to sanction. Follows ICC guidelines.
Produced a Green Claims code to tackle greenwash advertising. Produced a code with the automobile industry. Role to regulate sugar in soda and bad fat in food.	Funded and staffed by advertising industry personnel. Not independent. Ideologically committed to a consumerism model. The codes are not efficient against the negative effects of advertising. They treat complaints a posteriori so sanctions are ineffective.
Advertisers have to be able to prove their claims. Juries of people who are not advertisers.	Funded and staffed by advertising industry personnel. Ideologically committed to a consumerism model.
Claims to reach decisions quickly, and offers training to the advertising industry to comply with codes.	Tendency to withdraw sensitive complaints following intimidation from corporate lawyers. Issues no fines and decisions are non-binding. The board only responds to text, not images. Pro-consumerism default position and ideologically biased to techno-fixes to solve the climate crisis.

example of a company called out repeatedly over many years for false environmental claims which tenaciously continues to make them. Regulators' decisions are, for the most part, non-binding and they do not, or rarely, administer fines to the advertisers found in breach of advertising codes. An investigation from Greenpeace Sweden found that during a period of ten years, Reklamombudsmannen had only issued a single fine against companies that used deceptive environmental messaging.[9]

In November 2021, following more than 250 public complaints filed against the company for vague and misleading environmental claims in its advertising, Swedish-Danish dairy industry Arla Foods was taken to court and found guilty, following action by the national Consumer Agency. It was ruled against for misleading use of the wording 'net zero climate impact'.[10] But that was only

Figure 7.4 Arla Swedish dairy milk bottle with a misleading 'Net Zero Climate Footprint' label. (New Weather Sweden)

Figure 7.5 Shell's 'Drive Carbon Neutral' advert spotted at a petrol station in October 2021. (Badvertising)

after several complaints made two years earlier about the company's advertising had gone nowhere.

Similar pressure was mounted by climate campaigners and law students on oil giant Shell for its promotion of carbon neutral petrol under the slogan 'Drive Carbon Neutral' encouraging customers to pay 1 cent extra when filling their petrol tanks.[11] After the Dutch self-regulatory body ruled the advert to be misleading, Shell appealed against the decision, and changed the wording to promise instead 'CO_2 compensation', but the Dutch authority ruled against it a second time.[12] Despite such rulings against misleading adverts, the sheer difficulty and scale of public campaigning needed to make them happen highlights the dysfunctional nature of the process for effectively addressing the issue of high-carbon and greenwash advertising. What is needed are regulators that are proactive.

Adverts that promote 'socially irresponsible' behaviour and consumption also get an easy ride in these countries. In 2018, the French environmental energy agency (Ademe) and several other environmental organisations challenged one of its advertising regulators, the Jury de Déontologie Publicitaire, for not calling-out adverts for computers that promoted overconsumption.[13] The adverts in question featured people commenting for example: 'My old computer still works but an accident can occur any time' or, with tongue in cheek, 'I don't need a bigger screen. Although ... my eyesight is getting worse.'

In Belgium in October 2020, 102 complaints were sent to the regulator, Juri d'Ethique Publicitair, (JEP), about an advert for the car maker Jeep for a large hybrid vehicle which, they claimed, was 'inspired by nature'. The advertising authority resolved to raise the issue with the Belgian car makers' federations, Febiac and Traxio, who simply committed to direct an information campaign at their members about the CO_2 emissions of their vehicles, and about environmental regulations. Another advert by Jeep, (see Figure 7.6), this time in Sweden in 2021, for a big, hybrid SUV, claimed to offer 'ecological mobility' led to official complaints to the the Swedish ombudsman.

Figure 7.6 'Ekologisk' in Swedish translates as both 'ecological' and 'organic' so Jeep is offering ecological and organic mobility with a hybrid vehicle still dependent on fossil fuels. (New Weather Sweden)

SO WHAT SHOULD WE DO ABOUT IT?

We need to stop promoting our own self-destruction with high-carbon advertising. To do that we need new laws, codes and regulators with the power to act. We have looming in our very recent past the precedent of banning tobacco advertising that shows how it can be done. That just needs applying to advertising by major polluters, like the fossil companies, airlines, SUV makers.

In order to be able to regulate effectively, regulators need more resources and proper independence from the advertising industry. This implies moving from what we have at the moment, which is an industry-funded system of 'self-regulation' to one with rules and teeth operated by a genuinely independent public authority. Instead of relying on the general public to complain, who might not have detailed knowledge of the grounds it is possible to complain on, regulators also need to be proactive. In much the same way that UK television and cinema advertising that is considered political or contentious with regard to public policy issues gets blocked before being seen by the public, ads that contravene the guidelines on greenwash and promoting behaviour damaging to the environment should be screened-out prior to publication in print and online. That way harm could be prevented before it happens.

> Many States have adopted laws, but commercial advertising remains mostly self-regulated. This situation is unsatisfactory, leading to poor overall implementation, gaps, inconsistencies and legal uncertainty for both the industry and the public, as well as a paucity of clear, transparent and efficient complaint mechanism.[14] (Farida Shaheed, Special Rapporteur of the UN Human Rights Council)

What in particular could improve the system in the UK? Below we outline a few suggestions. In the next chapter, we take a step back and look at some lessons from the past, such as those from the ban on tobacco ads. We also look at how to make change happen, where change is already happening, and take a glimpse at how much better it would be if we could declutter the world from harmful and simply excessive advertising.[15] But first, what steps can be taken to stop advertising from further warming the planet?

A 'TOBACCO STYLE' LAW TO PROHIBIT HIGH-CARBON ADVERTISING

The regulator's interpretation of its own codes and remits is narrow, currently limited to regulating misleading and socially irresponsible advertising, meaning that as long as products and services are legal, even if they are harmful, it tends to hold its hands up as if it can do nothing. Given the scale and influence of advertising for the most polluting sectors and the urgency of the climate crisis, this is no longer sufficient. We have argued that it is logically possible to interpret its codes differently.

The UK House of Lords, alongside international bodies like UNEP and the WHO, are calling for national legislation to ban fossil adverts. Using a similar legal process to the banning of tobacco advertising, adverts for oil companies, highly polluting SUVs and airlines should be prohibited on climate and public health grounds. While people may be free to use these products, creating further demand for these harmful products through advertising seriously compromises the meeting of climate targets.

MAKE THE ASA A TRULY INDEPENDENT BODY

First of all, to ensure impartiality of the ASA, it is important that the funding of the organisation remains at arm's length of the industry. To honour the ASA's public mandate, revenues should be generated mainly via general taxation.

Greater distance should also be kept between advertising industry professionals and the regulator in the decision-making process. As it stands, the ASA Council and the 'industry panels' include representatives from the advertising industry. Instead, we propose that these should be replaced with an independent public panel advising on advertising complaints. The recruitment process should be open to members of the public and it should seek to have representatives from public health, environmental and consumer rights organisations to ensure a balanced view of interests.

This independent body should develop a body of rules informed by public policies in place to regulate advertising content prior to being broadcasted. This would ensure a far-reaching regulation of advertising content which, where necessary, could also rely on guidance from other regulatory institutions governing different sectors of the economy.

A MANDATE TO ISSUE LEGALLY BINDING RULINGS AND FINES

If the regulatory system was a building, it would be condemned as unfit for habitation. Regulation has an architecture all of its own and, at the moment, it is like a row of houses where several are simply missing rooms, a kitchen missing in one, a bathroom or bedroom in another. All arms of advertising's regulatory architecture need enforcement mechanisms including the ability to issue constraining sanctions and fines.

In order for the ASA's decisions to have a positive effect on companies' advertising practices, it is necessary that the organisation is given both the enforcement power to issue legally binding rulings and administer fines to advertisers in breach of regulation, and that it has sufficient resources to do more than conduct a small number of investigations in the hope that rulings will have a deterrent effect on others.

ENFORCEMENT POWERS TO SUSPEND AN ADVERT PENDING INVESTIGATION

In our experience complaints too often get brushed aside because the ASA is busy reviewing some policy or other, or is tied up in a related investigation. These two factors stop what should be normal, day-to-day complaint processes. We think that where there is reasonable cause to believe a complained-about advert contravenes the codes, it should be suspended until there is time and capacity for a deeper investigation.

Because of the sheer volume of greenwashing adverts, and the lack of resources to tackle the problem effectively, a basic 'greenwashing test' could be run for all advertising complaints against the ASA's Green Claims guidance and codes to suspend adverts prior to an investigation.

GREATER TRANSPARENCY AND EFFECTIVENESS IN THE COMPLAINT PROCESS

As we've seen, many of the advertising complaints sent to the ASA end up being discounted for various reasons that often stop well short of giving a reasoned explanation. The system needs much greater transparency, with clear, reasoned and evidence-based judgements on why or why not an investigation has gone ahead. For the sake of public accountability, a detailed breakdown of how complaints are categorised could also be made public.

To avoid significant delays between the filing of a complaint and its resolution, a reasonable time limit should also be respected. Past this deadline, the advert should be suspended until a final decision is made. Also, those who actually make the complaints should be regularly updated on the progress of their complaint.

CLARITY ON WHO DOES WHAT ACROSS ALL THE DIFFERENT REGULATORY BODIES

Greater clarity on the specific roles of every authority responsible for advertising regulation in the UK is urgently needed. Too often people trying to make a complaint get passed around with no one taking responsibility. There is a lack of easily available information on the specific mandates, and where they overlap, of all the different organisations concerned with advertising regulation in the UK. For instance, alongside the ASA, the Competition and Markets Authority, Trading Standards and the Financial Conduct Authority all have additional mandates to ensure that advertising aligns respectively with consumer protection and financial regulation.

One of the fundamental issues that we and others keep encountering, those who want a regulatory system for advertising that works and is in touch with the times, is that the ASA is not remotely set up to be a proper day-to-day working regulator in the way that it presents itself – as being there to police any advert that is wrong. What we mean by that is that there is an expectation and a duty, for example, on the actual police, to police all crime. But even though the ASA presents itself as the 'always there' watchdog, it doesn't even begin to operate in a proper policing function – its model is entirely different. With very little transparency its modus is simply to pick a few, demonstrative examples for occasional investigation (as we've seen, a tiny minority of actual complaints) and then rely on a handful of rulings to have a deterrent effect on transgressions in the rest of the market.

They have almost said as much in their correspondence with us, with investigations sometimes rejected, not on their merits, but because other 'similar' complaints against banks, or airlines or fossil fuel companies were being looked into. There is often no clear or public rationale for what does or does not actually get investigated. Or complaints might be turned away because the organisation is considering its approach to the aviation sector, or green issues more generally, even though current codes would apply. On one such occasion when we made a complaint to the ASA about the fossil fuel company Shell, it was rejected because the watchdog was in the middle of a 'project', but we were told that our submission would give them 'useful intelligence'.

It's a bit like the police saying, 'Thank you for letting us know your neighbour has been murdered, we won't be investigating, either because we are currently investigating another murder, or we are currently in the middle of a project about murder, but what you've told us will help us with our other enquiries and our project.' But, if you're an advertising watchdog, surely an advert is either wrong and breaks the codes or it does not, and if it does, it is the job of the watchdog to act.

We believe that the changes outlined further above will improve matters, but there are other more fundamental issues to address, which will be the focus of the next chapter.

8
A World Without Advertising

> You already know enough. So do I. It is not knowledge we lack. What is missing is the courage to understand what we know and to draw conclusions.
>
> —Sven Lindqvist, author of *Advertising is Lethal*, 1957

> The Government should introduce measures to regulate advertising of high-carbon and environmentally damaging products.
>
> —UK House of Lords Environment Committee, report into behaviour change, October 2022[1]

Preventing the loss of a habitable climate is being undermined by advertising. The problem is one of mixed messages. In one ear society hears the cautious, reasoned and yet still terrifying scientific warnings, and into the other are poured the emollient words of advertisers enticing us to keep on overconsuming – of course you need that 2-tonne SUV to go shopping, and another long haul holiday to a beach soon to disappear beneath the rising waves of a warming world.

We hear a lot about technological fixes that promise, one day, to save us. But we hear very little about the simple cultural and social things we can do to stop making things worse. An end to the way advertising actively promotes our own self-destruction through overconsumption would be a start. Even logical actions like this may seem exotic and unlikely to some when they first come across them. That was the case with the proposal to end tobacco advertising which now, in some countries and industries at least, is taken for granted. But we've seen how those gains were won in the face of fierce industry opposition, campaigns of misinformation against the ban, and political weakness and foot dragging. Nevertheless, progress was made.

There is less time for the cultural shift needed to move rapidly away from advertising-fuelled overconsumption by what is still a relatively wealthy, global minority. But, as we've explored in the preceding chapters, there are also many other things to be gained from taking action – as if preventing climate breakdown is not enough. Less stress, greater wellbeing, happier children who are more able to learn, engage and build friendships, towns and cities that are less visually polluted, and freedom from manipulative and intrusive surveillance marketing: these are to name just a few of the other benefits.

At some level we know intuitively that beyond the point of meeting a range of basic human needs – food, shelter, company, culture, love – accumulating more 'stuff' doesn't deliver the satisfaction it promises. Rather, it becomes a treadmill of endless want and typically operates as a mask for other problems.[2] Adverts provoke and exploit our wants, and wanting more, as one newspaper columnist put it, often is the result of 'needing to feel safe or superior to others: accumulation as protection. At its heart is vulnerability. And unless you address the root cause, nothing is ever going to be enough.'[3]

In this sense, advertising works as a form of misdirection, pointing us to the wrong places to find a good life, but keeping us dissatisfied and thinking that if we just buy that next thing, our lives will get better. The advertising codes forbid adverts to be misleading, but in an important sense, that of the promotion of consumer lifestyles as a path to happiness and fulfilment, advertising as a whole industry is misleading.

Fortunately, the things that do improve our wellbeing once basic material needs are met are well known and understood. Across the already large and burgeoning literature, five themes emerge as core to raising overall wellbeing.[4] Importantly, they represent a possible double-win. Because these are things that do not require additional income, they do not have to cost you money, even if they require you to think and spend time differently. More than that, they might actually save you money because they have little, if anything, to do with shopping for consumer goods or joining the jet set. If advertising is a form of expensive misdirection, promising, falsely that if you buy extra stuff you will feel better, the sort of things that really do deliver more wellbeing also liberate financially, by freeing you from the prompt to purchase.

They are the very things that the economics of consumerism tends to limit and suppress through the work and spend cycle, the norms of the working week, towns designed for cars not people, and the inescapable brain pollution of advertising.

The first of the five is that we should find ways to connect with the people around us, with family, friends, colleagues and neighbours, at home, work, school and in our local communities. Ironically, during the lockdowns of the Covid-19 pandemic, many discovered new opportunities to connect, people cautiously managed their engagement with the world, but a combination of home working, shorter work weeks, being 'furloughed' and taking advantage of local public green spaces for walks and exercise actually created more time and space to reconnect. There was a flourishing of public art, both in honour of the health service, but also for its own sake that added a new sense of community.[5] It happened in other ways too in response to being distanced, people made more effort to connect via social media apps and online.

The second of the five ways is about being active, day-to-day, by going for a walk or run or just stepping outside. The idea is to walk, take the stairs, cycle, play a game, garden, have a dance, absolutely whatever works. The trick is to find a physical activity that you enjoy and fits well with your own ability and the shape of your day. Cutting down on car use is also a huge money saver, and as more people switch to walking, cycling and other non-car sets of wheels, town and city designs will shift away from the privileged position afforded wealthier car owners (it's easy to forget that those on the lowest incomes already own far fewer cars).

Next is developing the habit to take notice of the world. This really is a freebie, one that can be done with a shift of outlook. Like the concept of mindfulness, it involves being curious, or remarking on the unusual or beautiful, such as noticing something different about a friend, the changing seasons, or just the everyday – what are those pictures hanging on the wall where you're eating lunch? But it is also about becoming aware and appreciative of what matters to you, through consciously noticing what you and others are feeling, and reflecting on the things you do and what happens to you.

To keep learning is another foundation of wellbeing. This means trying something new without worrying too much whether or not you'll be the world's best at it. This will take time – but instead of the time you might spend window shopping in a shopping centre or online in the hope you'll find a product to make you feel fulfilled – learning something can deliver intrinsic satisfaction and means that you'll become more in yourself, rather than just owning more stuff. It may require some outlay, but with better personal returns, and there are multiple opportunities for free learning. For example, you might rediscover an old interest, sign up for a course, or be responsible for something different at work. It's the act of learning itself that makes the difference, from fixing a bike to crushing fresh spices to cook a curry for the first time, or learning to play the recorder. These are the things that can also turn back the tide of deskilling that results, for example, from the heavily advertised culture of plastic packed, processed supermarket ready meals. In a stroke you eat better, feel better and reduce waste. And there's nothing romantic or 'middle class' about it, it's simply how most of humanity, for most of time, has fed itself – passing on knowledge, survival skills and recipes between generations.

Lastly, if you give, you not only enhance your own wellbeing, you help create the conditions for the reciprocity that helps make more convivial, resilient and adaptive communities. What you give could be your time, attention, friendship, you might lend a skill to help someone out or make a thing to gift. It's not complicated. Do something for a friend or a stranger. Thank someone. Smile. Volunteer your time. Join a community group or one of the mutual aid groups that sprang up during the Covid-19 pandemic, met a need and is still going. Look out, as well as in. It helps to complete the wellbeing cycle by connecting to others. Imagine you notice broken guttering on your elderly neighbour's house and offer to fix it. After some quick homework, a job done, tea, biscuits and a conversation full of neighbourhood stories later, both you and your neighbour will feel better. You'll have learned something too, and got to know someone.

There they are: connect, be active, take notice, keep learning, give. It sounds almost too simple. A key thing, easily overlooked, is that at whatever level of income you have, engaging with these approaches is first likely to save you money, easing the cost of living. These are also many of the foundations of the resurgence of mutual aid groups, which themselves echo the working-class economic and cultural solidarity of the early cooperative movement – that had not just the provision of affordable food and decent working conditions as an objective, but economic self-education too.[6]

They represent a different kind of economy. Advertising, on the one hand, promotes the self-seeking, competitive individualism of consumer capitalism – an economy that weirdly thrives and feeds itself on the unhappiness and loneliness it grows. A wellbeing, less financialised economy, built more on sharing and resilience, on the other hand, works by rebuilding community and more effectively meeting human needs.

So in this last chapter we take a look at how to make a world with less advertising, and where that world is already emerging. It includes a whirlwind tour of how some of the lessons from the successful campaigns against tobacco advertising might be adapted and applied to advertising by major polluters. But a world with at least much less such advertising (if it needs spelling out, we don't see a problem with, say, a grocery store advertising carrots and potatoes for sale at a certain price), and no advertising of products that amount to the promotion of our own ultimate self-destruction, raises other questions worth a look.

With the green economy of renewable energy, new companies in the circular economy, and new models based on repairing, sharing, reusing and recycling there should be no shortage of new advertisers able to step up to replace the polluters, just as others stepped in when tobacco was ruled out. But there are other ways forward too, by choosing advertising-free social media and online platforms, and by developing and expanding the economic models they use.

Then there are those places – some we've already visited – where towns, cities and transport networks are shaking off the shackles of polluter's adverts with many going further and cutting down more widely on the visual intrusion of advertising, especially in public places where it is impossible to actively give your consent to view it.

WE KNOW CHANGE CAN HAPPEN

'How do you sell death?' was the rhetorical question the tobacco industry dealt with. It turned out they had many ways. They also had many ways to resist attempts to stop them selling death. 'How do you sell extinction?' might be the contemporary equivalent for those industries marketing high-carbon products and lifestyles. Fortunately, there are lessons from the successful campaigns to control and end tobacco advertising that today's campaigns attempting to halt the promotion of our self-destruction can learn from. Here are a few:

- *The clarity of message and messenger matters.* For most of the 40-year campaign, messaging came either from radicalised doctors or – in the UK at least – from ASH, the highly effective heart of the anti-smoking campaign. A parallel for the climate emergency might be greater engagement of scientists who have understood the urgency of the crisis in public debates about the scale of necessary change. We know that there are several laudable examples, but having more endorse the kind of cultural step change needed to support sufficient system and behaviour change would help. A striking example of this is set by the Canadian Association of Physicians for the Environment (CAPE), which started the first initiative led by medical professionals pushing for an end to 'fossil ads'. Their campaign, 'Fossil Fuel Ads Make us Sick', was launched on Clean Air Day in 2022. An open letter signed by over 35 organisations representing over 700,000 health professionals in support of an ad ban was sent to government ministers.[7]

- *Get on top of the facts but don't assume that they alone will win the argument.* The failure to pin down the other side's research lengthened the campaign. But it may have been the dirty tricks of the tobacco lobby that ultimately made them unsupportable politically. Anti-smoking campaigners also had to win public arguments around what would happen if the ban was passed. There will always be side-effects to consider, such as where advertising money might go if the option to advertise high-carbon lifestyles is curtailed. Think, for example, of the curious return and ubiquity of smoking as a trope in television and film mentioned in Chapter 3 since laws prohibited tobacco advertising.
- *Building useful alliances with respected professionals.* There is no doubt that the involvement of grassroots doctors as campaigners, who could not simply be dismissed as against the system, was important. Following in their footsteps means allying similarly with expert professionals who are vocationally driven to work in the public good. Major social issues from air pollution to climate change already wash up at the doors of health and medical professionals. They are well placed to speak out on these impacts too.
- *Yet also respecting other people who want to take part.* Confronting people with uncomfortable truths and seeking consensual approaches are both almost always important parts of successful campaigns. But they can lead to tensions. It's important to accept that multiple strategies with different tactics are often needed. The anti-smoking campaign managed to accommodate not just peers of the realm, but also the BUGA-UP creative direct action campaigners in Australia and their amazingly imaginative spray-cans.
- *Use humour.* Nor was it just what the BUGA-UP campaigners achieved, or the Adbusters network in Canada. These are matched by the subvertisers and brandalists of today who take on fossil fuel companies, car makers and airlines. It was the

sheer power of the 'Butt of the Month' awards announced by ASH. These were able to force tobacco companies to withdraw new advertising campaigns, or even occasionally new products, because they had been made to look ridiculous. Some car adverts are begging to be made to look stupid.

- *Stay positive and campaign against the sin and not the sinner.* The BMA's campaign always emphasised that individual smokers were victims and not their target. Although, of course, the tobacco promoters and lobbyists were a different matter. On being positive, the BMA's first newspaper had a huge picture of the sky and the headline 'BREATHE!' Some words lift your spirits and some seem to close you down. Campaigners need the right language to help them achieve the best effect.
- *Understand when politicians can act – and how they might.* We need to remember that, most of the time, politicians generally only move after or in tandem with the public mood; you have to be able to help them take the right decisions.
- *Give local politicians something to do.* Both Bristol and Liverpool had their own smoke-free charters, and eventually so did eight other cities. In Canada, the first public authority to ban adverts was the Toronto Transit Authority. Councillors want to improve their cities and they are powerful potential allies. Below we'll see where the cities of today are starting to rule out advertising by major polluters.
- *Think ahead.* This seems to be one of the lessons that some of the most effective anti-smoking campaigners learned – to frame the situation positively, ready for the next appalling scandal, storm or forest fire.
- *Keep it simple.* Confusion over evidence is easy for oppositions to stir up, especially when dealing with bad faith industries who have a vested interest in preventing progress. It doesn't matter how complex aspects of the argument become, make sure the basic message is simple and that it speaks to people's best instincts.

WE KNOW CHANGE IS HAPPENING

Before becoming American vice-president in 1920, and president himself in 1923, Calvin Coolidge was the governor of Massachusetts from 1918 to 1920 and implemented a number of socially progressive measures. He backed public employees for a cost-of-living pay rise, and women and children had their working week limited to 48 hours – an earlier step in the ongoing movement for shorter working weeks and better life balance. But, largely forgotten to history, he also put limits on outdoor advertising.[8]

Now, there are bans on billboard advertising in at least four American states – Maine, Vermont, Hawaii and Alaska. The benefits surprised many. In the two years after Vermont had its last billboard removed, revenue from tourism rose by 50 per cent. Whether or not a direct consequence, removal clearly didn't harm Vermont's attractiveness to visitors. Two further states, Rhode Island and Oregon, banned new billboards.

Further controls are promoted by a campaign called Scenic America, that says: 'Visual pollution. Sky Trash. Litter on a stick. The junk mail of the American highway. Nothing destroys the distinctive character of our communities and the natural beauty of our countryside more rapidly than uncontrolled signs and billboards.' Scenic America gives over 700 examples of communities prohibiting the erection of new billboards on the grounds that 'billboard control improves community character and quality of life – both of which directly impact local economies'.[9]

The before-and-after effects of restricting billboards were studied in the towns of Williamsburg in Virginia, Houston in Texas and Raleigh in North Carolina. In each case, following the introduction of stricter controls, local businesses saw their trade increase. The implication is that the aesthetic appeal of the areas had improved, making them more attractive tourist destinations.[10]

We began with the pioneering story of how the Brazilian city of São Paulo acted to cut the visual pollution of public advertising, and how this led to other changes such as revealing the need to improve the conditions of low-income neighbourhoods that had been hidden behind billboards. It's worth remembering that the city's conservative mayor was acting against advertising per se and their impact on mental health as well as their intrusive domination of streetscapes.

The logic here, and that pursued by campaigns like Adfree Cities in the UK, Reclame Fossielvrij in the Netherlands, Werbefrei in Germany, Résistance à l'Agression Publicitaire in France, Proyecto Squatters in Argentina or Democratic Media Please in Australia is that public advertising removes your choice to 'not' view adverts.

At least in theory when watching films or going online some degree of choice can be exerted; there are tools to block online adverts, and advertising free channels or streaming services. But on the street or public transport networks you have not given your consent to be confronted with the advertising that bombards you. And this 'non-consensual' advertising is an affront to our ability to exercise choice over what we want to be exposed to.

But since São Paulo's initiative, momentum has been growing rapidly in cities around the world to curtail more specifically advertising by major polluters. In the northern hemisphere the Netherlands has been especially active. The municipality of Amsterdam became the first city in the world to ban high-carbon ads. Not only did the city pass a motion to stop fossil ads in its own jurisdiction, it also passed one calling on the national government to follow suit. An agreement was reached with the metro transport operator to end adverts for fossil fuel companies, fossil fuel-powered cars and budget airlines.[11]

A similar move was made by the Dutch city of Haarlem where they went one step further and included in the list of adverts to be controlled meat from industrial agriculture, which has a particularly high-carbon footprint. The ban was set to be implemented for new contracts from 2024,[12] and the municipality of Nijmegen passed an almost identical ban.[13] Action has also been taken in the Netherlands in cities from Amersfoort to Utrecht, Leiden and Enschede.[14]

North Holland then became the first province to ban high-carbon ads and included in their scope not just adverts for meat, but fish too. While in The Hague a motion was passed to ban polluting ads from bus shelters but, to begin with, not more broadly. Then a coalition agreement for the period 2022–26 adopted a commitment to end such ads.[15]

On the other side of the world, one of Australia's biggest cities, Sydney, passed a motion to take steps to end advertising by coal, petroleum and 'natural' gas companies. It's estimated that ads in the city reach 2 million people per day, making it a major promotional platform. The city's move was sparked by an open letter from over 200 health organisations and professionals, asking for the ban due to huge health and climate impacts of burning coal, oil and gas.[16] Sydney also called on the federal government to compensate community groups to enable them to drop fossil fuel sponsors.

At the time of writing, multiple other moves have been made to take bans forward in Australia, including by the state of New South Wales, and the municipalities of Fremantle, Maribyrnong, Moreland, Darebin, Yarra and Inner West. Councils in Wingecarribee, Glen Eira and Byron Bay have moved to end sponsorships by fossil fuel companies.[17]

These moves are significant in that they are happening in countries that are, or have been, home to major producers of fossil fuels, but things are happening elsewhere too. In Sweden, its two major cities, Stockholm and Gothenburg, are moving to end fossil advertising.[18] While in the city of Lund, a motion to end adverts for air travel has been proposed.[19] In neighbouring Finland, the country's oldest city, Turku, is developing climate criteria for outdoor advertising.[20]

With little fanfare, things have started moving in the UK too. In June 2021 Norwich City Council unanimously voted to adopt a motion in support of an ethical advertising policy for the city that ruled out environmentally damaging products.[21] Earlier the same year the major city of Liverpool introduced a low-carbon advertising policy that would prohibit adverts for polluting cars, SUVs, airline flights and fossil fuel companies. It was passed unanimously.[22]

Also in 2021, North Somerset Council introduced a 'Low Carbon Advertising Policy' on the advertising sites it manages.[23] Equally without drama, Cambridgeshire County Council introduced a policy that simply spelt out that permitted advertising had to be in line with the council's policies in other areas.[24] It said that the council 'does not consider the following companies, partnerships, organisations or individuals as suitable for entering into advertising or sponsorship agreements with', then listed these as, among others, 'those involved in the manufacture, distribution or wholesaling of tobacco-related products, alcohol, fossil fuels, pornography or addictive drugs'.

The policy went on to also specify a presumption against products and services that are 'injurious' to public health and bad for healthy lifestyles, and 'those whose business activities/practices do not align with the Council's wider values, corporate objectives and strategic goals, such as the environment and carbon accounting'. In other words, the council has asserted its right to say no to high-carbon advertising, because doing so is in line with its overall objectives to promote the public interest. These are just some examples of pioneering towns and cities that are taking action to stop adverts fuelling climate breakdown. Now that the idea is out there, more are emerging all the time. As we were writing this, we heard news that the UK towns of Basingstoke and Coventry had both also introduced these measures. Privately, however, some councillors with limited resources might be concerned about the threat of legal challenges. So, to reinforce others thinking of taking similar action, but unsure of their grounds we commissioned a legal opinion that demonstrated councils in the UK are well within their rights to do so.[25]

Richard Wald KC of Essex Chambers was asked to produce a legal opinion that could provide advice to local councils wanting to screen out high-carbon ads from sites they control. In particular, the review sought to assess two main points: any legal risks attached to a policy to end high-carbon advertising; and the design of a lawful policy to control high-carbon advertising.

It concluded that local authorities' taking action are well within their rights, legal powers and discretion to do so. There is, in fact, a strong legislative foundation to act, given the need set in UK legislation to reach net zero carbon emissions. Even more specifically, the

UK's latest carbon budget explicitly recognises the need to reduce demand for high-carbon activities. Wald also found that the risks of adopting bans are limited and the prospects of successful challenge by advertisers low.

WHAT GROWS IN THE SPACES LEFT BY ADVERTS

When the great political scientist Susan George was once asked what she would replace orthodox economic growth with, which in her view was a false indicator of progress and socially and environmentally destructive, she replied that it was like asking what you would replace cancer with. In other words she'd replace it with nothing, its absence would be sufficient.

Some raise the question about advertising. If it was removed or there was a lot less of it, what would replace it? And a similar answer might suffice. From what we know about the effect of advertising on human wellbeing, mental health, its visual pollution and how it drives destructive overconsumption, simply having less of it would be progress. But, in fact, there are places where interesting things are happening when it does get removed.

The city of Bristol in the UK which gave birth to the Adfree Cities network has been a pioneer in replacing billboard advertising with community arts projects. Campaigners there saw an opportunity for cities to be redesigned for purposes a world away from excessive consumerism towards initiatives that put social cohesion, creative pursuits and developing other, more satisfying human abilities at their core.[26]

Bristol has a rich multicultural heritage and strong independent streak which has been demonstrated in the way local people have 'repurposed' outdoor advertising sites for vibrant community art projects. The Burg Arts Project is focused on a repurposed billboard in the St Werburghs neighbourhood in East Bristol. Beginning in the years 2010–13, it showcased artworks by local artists. Then in 2018, local anti-advertising group, Adblock Bristol, worked with the St Werburghs Neighbourhood Association and St Werburghs City Farm to relaunch the project. The local area has seen a groundswell of resistance against corporate advertising, where residents have campaigned successfully to remove six out of 13 billboards.[27] A petition organised in the heart of the community showed 93 per cent of people in favour of all billboards being removed. In the five years since the group Adblock Bristol formed, 2017–22, they worked with residents all over the city to block more than 40 new digital advertising screens in the five years since the group's inception.[28]

Since 2006, in another area of the city, Stokes Croft, an outdoor art project has given artists the opportunity to exhibit on a privately owned wall, now part of a creative phenomenon that draws visitors from well beyond the city, known as the People's Republic of Stokes Croft (PRSC) 'Outdoor Gallery'. More than just an aesthetic wonder in its own right to visually improve the local environment, which would be enough, the initiative addresses how art can be a catalyst to discuss issues affecting the local and wider community, including climate change, social justice and the influence of corporate advertising. Both local and international artists have contributed, and the project has now given birth to an annual 'School of Activism'.

What is happening in Bristol is the result of well-planned local organising and a culture of street art, but nature abhors a vacuum and in the genteel and conservative Swiss city of Geneva, a legal stand-off between outdoor advertising contractors left a gap that was quickly and creatively filled by local people. Late in 2016, city officials invited companies to tender for a new outdoor advertising concession. When two bidding contractors became locked in a dispute there was a period of around three weeks when the billboards on the city's street were left blank, rid of commercial advertising. But they didn't stay bare for long. Residents quickly appropriated them, unleashing their creativity by filling the spaces with art and comment.

But it wasn't left there, the experience inspired residents to organise an ongoing campaign for a full ban on outdoor advertising.[29] During the Covid-19 pandemic, otherwise confined to their homes, many people rediscovered the value and joy of urban green spaces and nature. But before then, in 2015, Eric Piolle, the mayor of Grenoble, in France, and a member of the Green Party, had a novel approach to reconnecting people with nature. He ended an advertising contract with the hoardings company JC Deceaux and removed 326 outdoor commercial billboards, replacing them with a mix of trees and community noticeboards. 'It's time to move forward in making Grenoble a more gentle and creative city', said Piolle, who was elected the year before on a promise to clear advertising from the city. 'We want a city which is less aggressive and less stressful to live in, that can carve out its own identity. Freeing Grenoble of advertising billboards – "unbranding" the city streets – is a step in this direction.'[30]

Piolle's move would have pleased the poet Ogden Nash, who once sardonically observed that you are unlikely ever to see a billboard that was as lovely as a tree, and that unless we see the billboards fall, we may never 'see a tree at all'.[31]

A slightly less official, but very participatory approach is taken by ZAP Games based in Brussels but inspiring action in cities around the world. ZAP Games describes itself as an 'action-subversion game', in which multiple teams take simultaneous action against advertising. This involves replacing adverts in public places with a mixture of artworks and nature.[32]

But, as we've seen, this is not something restricted to action in the Global North. As far back as 2007, the municipal government of Beijing began reducing ads by targeting billboards for luxury housing. 'Many use exaggerated terms that encourage luxury and self-indulgence which are beyond the reach of low-income groups and are therefore not conducive to harmony in the capital', the city's mayor, Wang Qishan, told the *Wall Street Journal*.[33] In India cities ranging from New Delhi to Mumbai and Chennai introduced restrictions on outdoor advertising. Chennai banned new billboards in 2009. In Mumbai, a local group called Chal Rang De (Let's Go Paint) stepped in to show that in poorer neighbourhoods where houses are made from corrugated iron, painting with bright colours can help enhance, celebrate and demonstrate pride in community. In another extraordinary intervention in 2015, in the city of Tehran in Iran, the mayor organised a project called 'A Gallery as Big as a Town'. For ten days commercial billboards that were normally covered in calls to consume or religious slogans displayed art instead. The images ranged from Picasso to traditional Persian miniatures. 'It's pretty exciting', said Sadra Mohaqeq, an Iranian journalist with the reformist *Shargh* daily newspaper, 'For 10 days, people have time off from the usual billboard ads just promoting consumerism. It is going to affect people's visual taste in a positive manner.'[34]

On the theme of art replacing absent advertising in towns and cities, another case of it being better to have something simply missing is the wave of major artistic institutions that have dropped fossil fuel sponsors. In the UK, the Tate Gallery, National Portrait Gallery, Royal Shakespeare Company, Royal Opera House and Edinburgh International Festival have all chosen to step away from sponsorship arrangements with fossil fuel companies such as BP. In the Netherlands, the prestigious Van Gogh Museum is joined by the Science Museum Boerhaave and the Museum Museon (another science museum) in dropping sponsorship from oil company Shell. In Australia, the Questacon Science Museum also cut ties with Shell.

The ongoing experience of controlling the promotion of tobacco and addictive, nicotine-based cigarette substitutes shows what is possible, but also reminds that struggles like this require relentless vigilance. For that we are going to need regulators that regulate, as well as a vibrant civil society prepared to challenge irresponsible advertising. We've shown that already councils are well within their powers to refuse adverts from major polluters on sites they control. But as we explored in the previous chapter, the approach of regulators is dysfunctional: reactive, slow, under-resourced, under-powered, occasional, sporadic and 'after the event'. A high-carbon filter is needed that operates a 'presumption against' adverts that contradict the fact and the spirit of the UK's climate goals, and the officially recognised need to reduce demand for high-carbon activities. In other words, some very basic joining of the dots between already stated aims and applying fully language already contained in codes about adverts not encouraging environmentally damaging behaviour.

AD-FREE MEDIA – CUTTING DEPENDENCE ON ADVERTISING

If you want the media that you are able to choose to be ad free, there are already a number of ways to achieve that. Unlike with outdoor advertising that you cannot choose, and are forced to consume without your consent, as there is only one outdoors, take it or stay inside! In looking for new and better ways to do things, sometimes it's possible to overlook the hard-won public assets that are already available. If going 'ad free' sounds remotely radical, just look at a range of public service media platforms that already exist.

Of 31 such networks looked at in 25 countries by the Public Media Alliance, 13 excluded advertising and sponsorship funding altogether, including the UK's BBC, the powerful German broadcaster ARD and Japan's NHK. The public media of all Scandinavian countries plus the ABC and SBS in Australia did likewise.

Elsewhere, five public media networks in four countries – including Germany's ZDF, Spain's RTVE and Portugal's RTP – get less than 5 per cent of their funding from advertising. RTVE has refused all advertising since 2009, although some programmes may carry sponsorship.[35] Slightly less attractive, but an option that many take up, is to pay extra on some media platforms and streaming services to go advertising free. Other paid-for services are ad free to begin with. There are also 'ad-blocking' programmes you can upload so that when browsing online, adverts are filtered out, but sometimes this comes at the cost of access to the website you might be trying to view.

There are other alternatives in the online world specifically designed to filter out ads. The search engine DuckDuckGo, for example, makes a virtue of filtering out surveillance advertising. When the billionaire Elon Musk wreaked havoc on Twitter with his swaggering takeover of the micro-blogging and messaging site, users fled in herds to an alternative, Mastodon, that looks very similar, but is a not-for-profit, open source, decentralised and also ad free. So, there is already an emerging online ecosystem that means people can minimise their exposure to advertising.

Those are just some examples of how things are already done differently, but imagine how much more different, and better things could be with a greater range of media that are not in hoc to advertising, and for that matter, not controlled by a handful of billionaires with political agendas that are more concerned with preserving their wealth and influence than life on Earth. In practice, that might mean a range of things, from boosting the branches of public service media already in existence and finding ways to improve them, to experimenting with more open, democratic and local media, and looking into different media economic models.

THE ECONOMICS OF LOOSENING ADVERTISING'S GRIP

When companies spend money on promoting their brand, their corporate image, as opposed to just advertising their products for sale, it gets written off as a tax deductible cost. The logic is that brand promotion is a long-term investment rather than an operational cost and as such should be treated differently in tax terms. The failure to do so deprives public coffers of tax revenue. It's the difference between investments and expenses. Being able to write off large amounts of brand-promoting advertising against tax increases its attractiveness and leads to a lot more of it. It also gives unfair advantages to big businesses who have more to spend, and hence reinforces the monopoly power (technically 'oligopoly' power) of giant brands. There is a model being looked at in the United States that would substantially separate the two, making spending on brand advertising less tax advantageous and so less attractive.[36]

From an economic point of view, spending huge amounts on brand promotion is also inefficient, being an obstacle to both innovation and fair competition. Take for example the cosmetics giant L'Oréal who spent the equivalent of 29 per cent of its sales on commercial communication in 2015 (€7.4 billion) compared to less than

3 per cent on research and development.[37] Not only does this have a chilling effect on ensuring markets are open for smaller players or start-ups who cannot remotely match such spending power – large size is a bias itself that creates unfair competition – but the ability to spend vast amounts on brand building is inefficient in another sense too.

By dominating the market of information about products available to consumers the big brands can charge higher prices than if the market was genuinely open and fully competitive. Brand-building for consumer recognition is part of this process. In this sense, large-scale spending on brand and corporate identity is what economists would call unproductive expenditure. It leads to poorer outcomes for consumers, the public interest and a more closed market, but gives large corporations the power to manipulate whole sectors. To stop this, regulators at the national or European level could insist on spending caps, a bit like the way that some sports apply spending caps to clubs in national sporting leagues to ensure more honest and even competition.

There is a long history of proposals to tax advertising, partly as a way of avoiding just such an unproductive arms race of spending by a handful of giant corporations intent on market domination. France, for example, levies a small tax on advertising, set at around 2.5 per cent. Given the arguments in this book, a higher rate would have not only a deterrent effect on 'over-advertising', but an income stream that could be directed at creating greater media balance by funding more democratic, public interest, community-based and citizen-led media. This is an idea that has been around for some time, but given current concerns around concentration of media ownership, and the dominance of commercially driven, billionaire-owned media, its time is ripe. Several highly influential, privately owned national media voices like the *Daily Mail* and *Daily Telegraph* consistently provide platforms that undermine climate science and question the need for action.

An advertising tax was suggested in theory by Hungarian economist, Nicholas Kaldor, as early as 1961, and subsequently applied in practice by Sweden in 1971 and France in 1982. More thinking on the subject was done in the 1990s by Edwin Baker, a US expert on economic democracy in the media, in his 'Tax Advertising – Subsidize Readers'. The proliferation, penetration and influence of digital advertising has renewed and sharpened interest in creative measures to tax advertising.

The idea that space for different views needs to be opened up and protected for democracy to function is a well-established principle. Many nations, for example, have laws around election time that seek to prevent the abuse of power by dominant and wealthy voices. Under the Broadcasting Act in the UK a system of broadly equitable allocation based on demonstrable support controls access to party political broadcasts on radio and television. But the idea that it is only the short span between an election being called and votes being cast is the only time when a semblance of democracy needs to be created and protected to guard against systemic bias belies the reality of concentrated power and control online, and in broadcast and print.

One option for greater mainstream diversity of voice and opinion would be the creation of new public media, applying principles that are standard during election times more broadly.

In *The Return of the Public*, Dan Hind concluded that media commissioning in general needs revolutionising. He argues that the media's ability to lend credence to false threats to the public and ignore or downplay others is a problem that requires innovation. The sectoral group-think that, for example, failed sufficiently to question the existence of weapons of mass destruction in Iraq led to a war, the consequences of which still haunt. Conversely the failure to question adequately, or at all, the systemic flaws in banking and finance, or report on the full implications of climate breakdown and the changes needed adequately to address it, is a glaring, gross and life-threatening professional failure. He writes that if the system that 'provides most people with most of their information about the world beyond their immediate experience' can solidify world-changing illusions, on the one hand, and miss unexploded bombs, on the other, we need new types of media. The international group Reporters Without Borders publish an annual index of press freedom around the world. Its 21st edition in 2023 looked at the 'radical changes' linked to 'political, social and technological upheavals'. What it found was a dire state of health for media around the world, with the conditions for journalism 'bad' in seven out of ten countries. The UK was picked out for threats to investigative journalism from 'a national security bill that lacks protective measures'.[38]

Hind wants a shift in the centre of gravity for the decisions that determine what we get to hear about and how. He argues for a type of informed, democratic, crowdsourced system of public commissioning that would allocate resources to researchers and investigative journalists. It would sit alongside existing, conventional media, but dilute their control. In practice this would widen the realm of civic engagement, allowing individuals otherwise excluded to raise matters of common concern to fellow citizens. One consequence of this would be to make other forms of civic participation seem less daunting or pointless. By giving the general population the means to enquire into issues, concerns and challenges, public commissioning could publicise the facts that might build foundations for political change and, for example, necessary climate action. The successful use of citizens assemblies to explore issues and decide on priorities for local action gives a glimpse of how such a process, funded for example by a tax on advertising, could operate. Another precedent, though very mild, can be seen in the UK government's e-petitions initiative.

Under a system of public commissioning a share of programming and investigative journalism would be determined democratically by public interest. It would be a genuinely civic alternative, the equivalent, as one of us has written before, 'of the public town square as opposed to the private, gated, shopping mall or state department store, the commons to stroll on instead of the themed leisure park'.[39] As a form of quality control, and to prevent being hijacked by vested interest, open-access media commissioning could be made to conform to a public interest test.

In 2012, an official investigation into long-standing criminal practices by tabloid newspapers in the UK shook the foundations of faith in mainstream journalism. The Leveson Inquiry had exposed the extraordinary lengths that papers were prepared to go to in selling copies, including hacking the phone of a murdered schoolchild. In response, The Bristol Cable was set up as a crowdfunded, reader-owned new media outlet 'to do something about the failures' of big corporate media. The focus on local investigative journalism and their model of being '100% owned by thousands of local people, free to access for all in digital and print', means they can engage in long-term investigations, rather than churning news 'that relies on clicks for revenue'. They are just one pioneering initiative, but local investigative journalism has a proud history, and they see themselves as locally rooted but part of a 'global movement to reinvent the media industry as a sustainable, democratic public service'.

Proper support and funding directed to new media operators like Bristol Cable that fulfil public interest and service criteria could begin to transform the media landscape and liberate us from the pervasive influence of advertising.

THE MINISTRY FOR THE CLIMATE EMERGENCY

But time is short for climate action. While we wait for the great transformation of a media system that, consciously or not, promotes self-destruction with adverts for high-carbon products and services we created our own government department, the Ministry for the Climate Emergency. Its sole purpose so far has been to produce light-hearted public information campaigns to counter the relentless advertising propaganda of major polluters.

In fact it is more than one ministry, we precociously set up equivalent government departments everywhere from the United States, to France, Sweden and beyond. You can see some of their work alerting the general public to the dangers of advertising's brain pollution below.

Research tells us that we make better decisions if we can visualise ourselves in the world in, say 20 years time.[40] By then the consequences of a whole range of life choices are likely to show, and perhaps we become a little clearer about what actions are needed to make sure it's a world we'd like to live in, and that we, ourselves, are the people we'd like to be. This last chapter has attempted a glimpse at a world with less advertising in it, and we think that would be a better place, with happier people and less pressure on the rest of nature.

We leave you with the Ministry's opening warning:

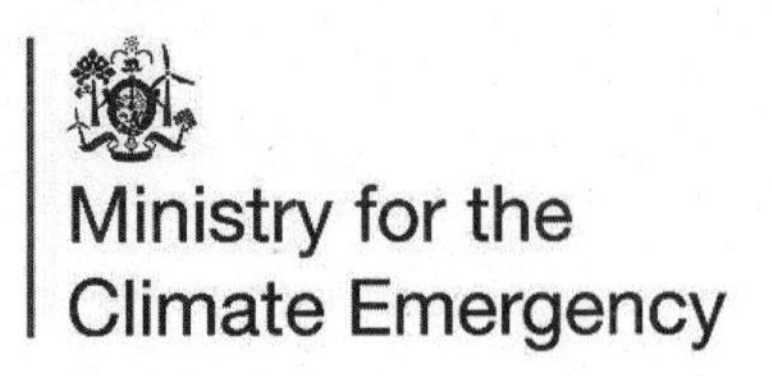

We all have Brain Pollution.
We must act now before it is too late
for us and the climate emergency.

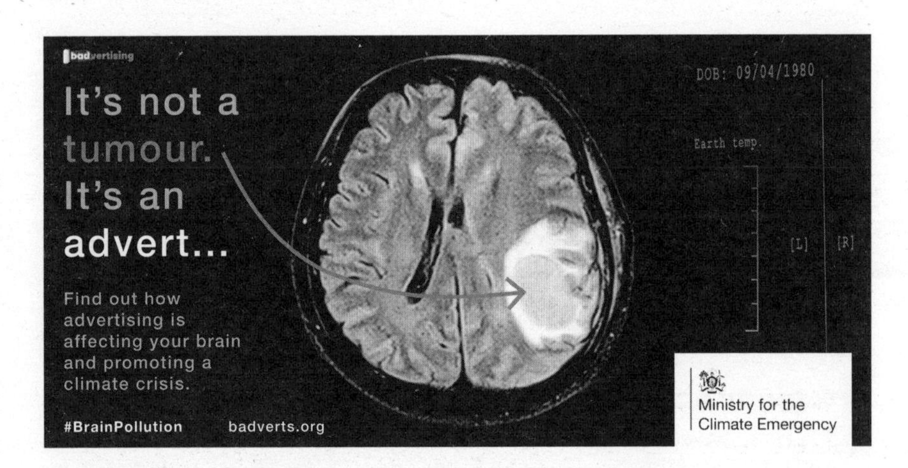

Figures 8.1–8.6 Public health and climate emergency information campaigns by the Ministry for the Climate Emergency.

You have
Brain Pollution.
Act now before
it's too late.
Find out how advertising
is promoting a climate crisis.
#BrainPollution badverts.org
Ministry for the
Climate Emergency

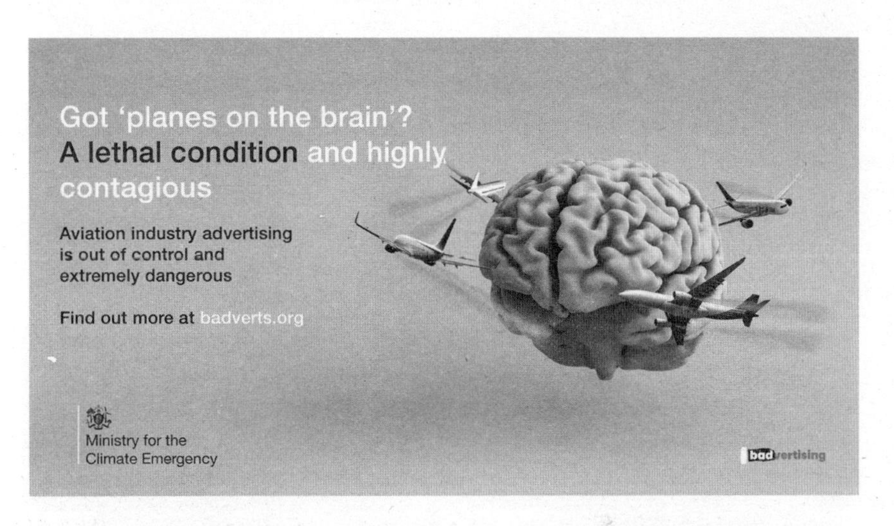
Got 'planes on the brain'?
A lethal condition and highly
contagious
Aviation industry advertising
is out of control and
extremely dangerous
Find out more at badverts.org
Ministry for the
Climate Emergency
badvertising

Grounded
It's time to cancel aviation
industry advertising before we
cancel life as we know it
Find out more at badverts.org
Ministry for the
Climate Emergency
Departures 23:59
VN521 ICE CAPS CANCELLED
PW605 CHILD'S FUTURE CANCELLED
RI770 CLEAN AIR CANCELLED
AY326 NATURE CANCELLED
TI472 FAIRNESS CANCELLED
CW833 WELL BEING CANCELLED
TR017 CARING CANCELLED
GA020 GREENWASH BOARDING
LT246 PLANET
badvertising

If that's an eco car
then I'm a unicorn.
Don't believe the carwash.
Find out how car advertising
is promoting a climate crisis
at badverts.org
badvertising
Ministry for the
Climate Emergency

Now with improved
anti flood features
Don't believe the carwash.
Find out how car advertising
is promoting a climate crisis
at badverts.org
badvertising
Ministry for the
Climate Emergency

Acknowledgements

As you will see throughout this book, over decades many voices other than our own have been raised against the corrosive effects of advertising, and the toxic cultures of materialism and consumerism that it fuels. We are grateful to all of those, from the earliest to the most recent critics of accumulation and overconsumption who point out that, if it's a 'good life' you want, the promises made by advertising are keys to an empty room.

But there are some very specific people and organisations that we must thank. First of all of these are our colleagues in the Badvertising campaign. We have been on a journey together and seen concern and action over adverts that fuel the climate emergency grow at an incredibly rapid rate. And it is people like our colleagues whose efforts shine through this book who are making this happen, including Emilie Tricarico, Robbie Gillett, Veronica Wignall, Freddie Daley, Anna Jonsson, Gunnar Lind, James Ward and Matt Bonner. The work of all of you has found its way into this book. A special thank you goes to David Boyle for help with research, especially concerning the history of the campaign to end tobacco advertising in the UK; and also to Jamie Beevor, our go-to number cruncher for all things carbon and transport. The brilliant creativity of Webster Wickham and his agency, BWA, has hugely enhanced our campaign.

Fellow travellers in the Badvertising initiative are the New Weather Institute, the climate action charity Possible, and Adfree Cities. We thank everyone in these organisations who have been part of raising an issue that is almost invisible because it is absolutely everywhere. The fact that the Badvertising campaign happened at all owes a huge debt of gratitude to the imagination and tenacity of Kate Power, then at the KR Foundation, who some time back, invited Andrew and Leo to convene a Commission of inquiry into the behavioural obstacles and opportunities to preventing planetary meltdown. One of the recommendations to emerge from the Commission's report became the Badvertising campaign. Since then, the campaign has been kindly supported by the KR Foundation, and encouraged by Brian Valbjørn Sørensen and Asger Narud, who we warmly thank too.

While advertising has been telling us for decades that the only limit to our consumption is our ability to pay – and they have ways around that too – there are many groups who have been marginalised as outlandish for having the audacity to suggest we should leave some space for the rest of nature, and point out that no spare, habitable planet is on hand once we've devoured this one. Networks like Subvertisers International, guerrilla 'Brandalists', old time Adbusters, and more, will one day be seen as the early warning system for a mainstream economics that failed to understand real human wellbeing and pushed people over a climate and ecological cliff in a misdirected search for happiness.

We are working with many other brilliant groups on a day-to-day basis to try and turn a cultural corner. Every day it seems they grow in number, but our regular collaborators include: New Weather Sweden, Reclame Fossielvrij, Résistance à l'Agression Publicitaire, Adblock Bristol, Bruxelles Sans Pub, Creatives for Climate, Fossil Free Media, Comms Declare, Werbefrei, Proyecto Squatters, and all the dynamic organisations in the the Cool Down network of sport for climate action, and many more. As a question of long-term survival, collectively they are working to help societies step back from consumerism's competitive, self-absorbed individualism, and step towards a more cooperative, collaborative world that actually thinks about how we may thrive within planetary limits by valuing people and nature more.

We'd also like to thank those people who are trying to raise awareness of the climate emergency within the advertising industry, agencies like Glimpse and networks like Purpose Disruptors, Clean Creatives and the Conscious Advertising Network. Many others within the industry and the organisations which oversee it hover beneath the radar but do good work trying to curb advertising's excesses. One thing which could bring change almost faster than anything else would be for more creative working within advertising to reject the briefs offered by heavily polluting clients.

Lastly, and in terms of this book finding its way into the public domain, we would like to say a very large thank you to David Castle, editor at Pluto, who took on the project at short notice, saw its potential, and has been a cheerleader for it since. Thanks also to Carolyn Russo, Curator of Art at Smithsonian's National Air and Space Museum.

Notes

INTRODUCTION: ADVERTISING AND THE INSIDIOUS RISE OF BRAIN POLLUTION

1. See: About Earth Overshoot Day. www.overshootday.org/about-earth-overshoot-day/
2. Climate Uncensored (2022) What's in a name? https://climateuncensored.com/whats-in-a-name/
3. Committee on Climate Change (2009) Meeting the UK aviation target – options for reducing emissions to 2050: Chairman's Foreword, December.
4. The White House (2023) FACT SHEET: President Biden's budget lowers energy costs, combats the climate crisis, and advances environmental justice. www.whitehouse.gov/omb/briefing-room/2023/03/09/fact-sheet-president-bidens-budget-lowers-energy-costs-combats-the-climate-crisis-and-advances-environmental-justice/
5. Ad Age (2003) History: 1970s. https://adage.com/article/adage-encyclopedia/history-1970s/98703
6. New York Times (2007) Anywhere the eye can see, it's likely to see an ad. https://web.archive.org/web/20200320191358/https:/www.nytimes.com/2007/01/15/business/media/15everywhere.html
7. Carr, Sam (2021) How many ads do we see a day in 2023? Lunio.ai, 25 February. https://lunio.ai/blog/strategy/how-many-ads-do-we-see-a-day
8. Statista (2023) Digital advertising – worldwide. www.statista.com/outlook/dmo/digital-advertising/worldwide.
9. See for instance: Possible (2022) Missed targets: a brief history of aviation climate targets, 9 May. www.wearepossible.org/our-reports-1/missed-target-a-brief-history-of-aviation-climate-targets
10. www.livingstreets.org.uk
11. Rapid Transition Alliance (2019) Adblocking – the global cities clearing streets of advertising to promote human and environmental health. www.rapidtransition.org/stories/adblocking-the-global-cities-clearing-streets-of-advertising-to-promote-human-and-environmental-health/
12. Environmental News Network (2007) 7 June. www.enn.com/articles/22788-sao-paulo-bans-outdoor-ads-in-fight-against-pollution
13. Lapierre, Matthew A., Frances Fleming-Milici, Esther Rozendaal, Anna R. McAlisterr and Jessica Castonguay (2017) The effect of advertising on children and adolescents. *Pediatrics*, 140(Supplement_2), November, S152–S156. https://doi.org/10.1542/peds.2016-1758V
14. See: https://commercialfreechildhood.org/

15. Kasser, Tim (2002) *The high price of materialism*. Cambridge, MA: MIT Press.
16. Badvertising (2020) *Upselling smoke*. https://static1.squarespace.com/static/5ebd0080238e863d04911b51/t/5f21659998148a15d80ba9be/1596024223673/Upselling+Smoke+FINAL+23+07+20.pdf
17. See: https://subvertisers-international.net/ and https://badvertising.se
18. See: https://antipub.org/
19. See: https://verbiedfossielereclame.nl/
20. See: www.canopea.be/
21. See: www.duh.de/projekte/verbrauchertaeuschung-in-der-autowerbung/

1 BADVERTISING, PRIMING, PROPAGANDA AND SURVEILLANCE ADVERTISING

1. EasyJet (2018) Imagine. www.youtube.com/watch?v=FoKOBPfJYSU
2. Bite Back 2030 (2020) The video the fast food industry don't want you to see. 18 January. www.biteback2030.com/news/watch-video-fast-food-industry-dont-want-you-see
3. Schaeffer, M. et al. (2006) Neural correlates of culturally familiar brands of car manufacturers. *NeuroImage*, 31(2), 861–5. https://doi.org/10.1016/j.neuroimage.2005.12.047
4. McClure, S.M., et al. (2004) Neural correlates of behavioral preference for culturally familiar drinks. *Neuron*, 44(2), 379–87. https://doi.org/10.1016/j.neuron.2004.09.019
5. Ambler, T., A. Ioannides and S. Rose (2003) Brands on the brain: neuro-images of advertising. *Business Strategy Review*, 11(3), 17–30. http://dx.doi.org/10.1111/1467-8616.00144
6. Greenbook, Hofmeyr J. (2015) How brands really grow 3: brands in the brain. www.greenbook.org/mr/market-research-news/how-brands-really-grow-3-brands-in-the-brain/
7. Patti, M. Valkenburg and Moniek Buijzen (2005) Identifying determinants of young children's brand awareness: television, parents, and peers. *Journal of Applied Developmental Psychology*, 26(4), 456–68.
8. Tatlow-Golden, Mimi, Eilis Hennessy, Moira Dean and Lynsey Hollywood (2014) Young children's food brand knowledge. Early development and associations with television viewing and parent's diet. *Appetite*, 80. www.sciencedirect.com/science/article/abs/pii/S0195666314002177
9. McAlister, Anna R. and T. Bettina Cornwell (2010) Children's brand symbolism understanding: links to theory of mind and executive functioning. *Psychology & Marketing*, 11 February. https://doi.org/10.1002/mar.20328
10. Fitzsimons, G.M., T.L. Chartrand and G.T. Fitzsimons (2008) Automatic effects of brand exposure on motivated behavior: how apple makes you 'think different'. *Journal of Consumer Research*, 35(1), 21–35. https://doi.org/10.1086/527269

11. Chan, H-Y., M. Boksem and A. Smidts (2018) Neural profiling of brands: mapping brand image in consumers' brains with visual templates. *Journal of Marketing Research*, 55(4), 600–15. https://doi.org/10.1509/jmr.17.0019
12. See: Kasser, Tim et al. (2020) A*dvertising's role in climate and ecological degradation*. London: Badvertising.
13. That is the conclusion of the next chapter. See: ibid. https://static1.squarespace.com/static/5ebd0080238e863d04911b51/t/5fbfcb1408845d09248d4e6e/1606404891491/Advertising%E2%80%99s+role+in+climate+and+ecological+degradation.pdf
14. Kearney, J. (2010) Food consumption trends and drivers. *Philosophical Transactions of the Royal Society, Biological Sciences*, London, 27 September. www.ncbi.nlm.nih.gov/pmc/articles/PMC2935122/#RSTB20100149C50
15. Putnam, J. and J.E. Allshouse (1999) Food consumption, prices and expenditures, 1970–1997. US Department of Agriculture Statistvical Bulletin 965. Washington, DC: US Department of Agriculture.
16. Willett, W. (2002) *Eat, drink and be healthy: the Harvard Medical School guide to healthy eating*. New York: Free Press.
17. Bargh, J.A., M. Chen and L. Burrows (1996) Automaticity of social behavior: direct effects of trait construct and stereotype activation on action. *Journal of Personality and Social Psychology*, 71(2), 230–44. https://doi.org/10.1037/0022-3514.71.2.230
18. Bauer, M.A., J.E. Wilkie, J.K. Kim and G.V. Bodenhausen (2012) Cuing consumerism: situational materialism undermines personal and social well-being. *Psychological Science*, 23(5), 1 May, 517–23.
19. Ibid.
20. Piff, P.K., D.M. Stancato, S. Côté, R. Mendoza-Denton and D. Keltner (2012) Higher social class predicts increased unethical behavior. *Proceedings of the National Academy of Sciences*, 109, 4086–91.
21. University Health News (2020) Is Red Bull bad for you? Why you should steer clear of energy drinks, 18 December. https://universityhealthnews.com/daily/energy-fatigue/is-red-bull-bad-for-you-4-reasons-to-skip-these-dangerous-drinks/
22. Centers for Disease Control and Prevention, Dangers of Mixing Alcohol and Caffeine. www.cdc.gov/alcohol/fact-sheets/caffeine-and-alcohol.htm. Quoting: Roemer, A. and T. Stockwell (2017) Alcohol mixed with energy drinks and risk of injury: a systematic review. J*ournal for the Study of Alcohol and Drugs*,78(2), 175–83.
23. Munteanu, C., C. Rosioru, C. Tarba and C. Lang (2018) Long-term consumption of energy drinks induces biochemical and ultrastructural alterations in the heart muscle. *The Anatolian Journal of Cardiology*, 9(5), May, 326–3. doi: 10.14744/AnatolJCardiol.2018.90094
24. Science Daily (2011) Energy drink logo enough to shape consumer performance, study finds, 31 January. Quoting: Brasel, S. Adam and James Gips (2011) Red Bull 'Gives You Wings' for better or worse: a double-edged impact of brand exposure on consumer performance. *Journal of Consumer*

Psychology, 21(1), Special Issue: Nonconscious Processes in Consumer Psychology, 57–64.
25. Ibid.
26. Ibid.
27. Matthews, Dawn (2020) Colour psychology: how colour meanings affect your brand. Avasam.com blog, 25 February. www.avasam.com/colour-psychology-how-colour-meanings-affect-your-brand/
28. Spence, C. (2015) On the psychological impact of food colour. *Flavour*, 4, 21. https://doi.org/10.1186/s13411-015-0031-3
29. Modern Restaurant Management (2022) Understanding the psychology of color in restaurant design. 10 July.
30. Zalis, Shelley (2019) Busting gender stereotypes: the pink versus blue phenomenon. Forbes.com, 5 September. www.forbes.com/sites/shelleyzalis/2019/09/05/busting-gender-stereotypes-the-pink-versus-blue-phenomenon/
31. Schauss, A. (1979) Tranquilizing effect of color reduces aggressive behavior and potential violence. *Journal of Orthomolecular Psychiatry*, 8, 218–21.
32. Elliot, AJ. (2015) Color and psychological functioning: a review of theoretical and empirical work. *Frontiers in Psychology*, 6, 2 April, 368. doi: 10.3389/fpsyg.2015.00368
33. See for example: Prochnow, D., H. Kossack, S. Brunheim, K. Müller, H.-J. Wittsack, H.-J. Markowitsch and R.J. Seitz (2013) Processing of subliminal facial expressions of emotion: a behavioural and fMRI study. *Social Neuroscience*, 8(5), 448–61. https://pubmed.ncbi.nlm.nih.gov/23869578/
34. Simion, F. and Di Giorgio, E. (2015) Face perception and processing in early infancy: inborn predispositions and developmental changes. *Frontiers in Psychology*, 6, 9 July. doi: 10.3389/fpsyg.2015.0096
35. Biles, Anthony (2020) 'How I created the Amazon logo', Brand Berries. 10 September. https://www.thebrandberries.com/2020/09/10/how-i-created-the-amazon-logo/
36. Fernandez, Jon (2010) Coca-Cola brings back Open Happiness push. *Marketing Week*, 8 February. www.marketingweek.com/coca-cola-brings-back-open-happiness-push/
37. Subliminal priming – state of the art and future perspectives. www.mdpi.com/2076-328X/8/6/54/htm
38. Rogers, Stuart (1992) How a publicity blitz created the myth of subliminal advertising. *Public Relations Quarterly*, 37, Winter, 12–17. www.subliminalworld.org/aaaa3.htm
39. O'Barr, William M. (2005) 'Subliminal' advertising. *Advertising & Society Review*, 6(4), 10 August. www.semanticscholar.org/paper/%22Subliminal%22-Advertising-O%E2%80%99Barr/323547fa99d6a9fcdc824ab11dfbc26f533a05c3
40. Karremans, Johan, Wolfgang Stroebe and Jasper Claus (2006) Beyond Vicary's fantasies: the impact of subliminal priming and brand choice. *Journal of Experimental Social Psychology*, 42, 792–8. www.researchgate.net/

publication/222416467_Beyond_Vicary's_fantasies_The_impact_of_subliminal_priming_and_brand_choice/citations

41. See for example: Greenwald, A.G., E.R. Spangenberg, A.R. Pratkanis and J. Eskenazy (1991) Double-blind tests of subliminal self-help audiotapes. *Psychological Science*, 2, 119–22. https://journals.sagepub.com/doi/10.1111/j.1467-9280.1991.tb00112.x. Audley, B.C., J.L. Mellett and P.M. Williams (1991) Self-improvement using subliminal audiotapes: consumer benefit or consumer fraud? Presented at the Meeting of the Western Psychological Association, San Francisco, CA, April. And finally: Merikle, P.M. and H.F. Skanes (1992) Subliminal self-help audiotapes: search for placebo effects. *Journal of Applied Psychology*, 77(5), October, 772–6. https://pubmed.ncbi.nlm.nih.gov/1429349
42. Elgendi, M., P. Kumar, S. Barbic, N. Howard, D. Abbott and A. Cichocki (2018) Subliminal priming – state of the art and future perspectives. *Behavioral Science* (Basel), 8(6), 30 May, 54. www.ncbi.nlm.nih.gov/pmc/articles/PMC6027235/
43. Science Daily (2011) Energy drink logo enough to shape consumer performance, study finds. 31 January.
44. Ibid.
45. Schindler, Michael (2022) The future of advertising: the next 10 years (updated). 10 December. https://voluum.com/blog/future-of-advertising/
46. Ibid.
47. Statista (2023) Meta: annual advertising revenue worldwide 2009–2022. 13 February. www.statista.com/statistics/271258/facebooks-advertising-revenue-worldwide/
48. Ortiz-Ospina, Esteban (2019) The rise of social media. Published online at https://ourworldindata.org/rise-of-social-media
49. Boyle, David (2004) *Authenticity: brands, fakes, spin and the lust for real life*. London: Harper Perennial, 266–7.
50. Bernays, Edward (1928) *Propaganda*, 2nd edn. New York: Horace Liveright, 73. https://archive.org/details/Propaganda_Edward_L_Bernays_1928.pdf
51. Ibid., 74.
52. Ewen, Stuart (1996), PR! A social history of spin. New York: Basic Books, 162–3.
53. Cutlip, Scott M. (1994) *The unseen power: public relations. A history*. Hove, UK: Lawrence Erlbaum, 168.
54. Bernays, Edward (1928) Manipulating public opinion. *American Journal of Sociology*, 33, May, 958–71.
55. See: Curtis, Adam, Century of self – happiness machines. www.youtube.com/watch?v=Mojw7DIpu1k
56. Lundberg, Ferdinand (1946) *America's sixty families.* New York: The Citadel Press, 313n.
57. Bernays, Edward (1965), *Biography of an idea: memoirs of public relations counsel.* New York: Simon & Schuster. https://archive.org/details/biographyofideamoobern

58. Bernays, Edward (1947) The engineering of consent. *Annals of the American Academy of Political and Social Science*, 250, March, 113. Reprinted in Bernays, Edward L. and Howard Walden Cutler (1955), *The engineering of consent*. Norman, OK: University of Oklahoma Press.
59. Ibid.
60. Ibid.
61. Bernays, Edward L. (1923) *Crystallizing public opinion*. New York: Boni and Liveright, 121–2. https://gutenberg.org/ebooks/61364
62. Walker, Stanley (1934) *City editor*. New York: Frederick A. Stokes Co., 145–6.
63. Branfied, Tracy (2022) The ads have eyes: is surveillance marketing worth the risk? Clockwork Media. www.clockworkmedia.co.za/the-ads-have-eyes-is-surveillance-marketing-worth-the-risk/
64. Vox (2018) The Facebook and Cambridge Analytica scandal explained. 2 May. www.vox.com/policy-and-politics/2018/3/23/17151916/facebook-cambridge-analytica-trump-diagram 7
65. Statista (2023) Number of daily active Facebook users worldwide as of 1st quarter 2023. www.statista.com/statistics/346167/facebook-global-dau/
66. The Washington Post (2020) Facebook content moderator details trauma that prompted fight for $52 million PTSD settlement. 13 May.
67. Time Magazine (2023) Facebook content moderators sue meta over layoffs in Kenya. 20 March.
68. Federal Trade Commission (2021) Press release. Washington, DC. 15 December. www.ftc.gov/news-events/news/press-releases/2021/12/advertising-platform-openx-will-pay-2-million-collecting-personal-information-children-violation
69. New York Times Magazine (2012) How companies learn your secrets. 19 February. www.nytimes.com/02/19/magazine/shopping-habits.html
70. Simms, Andrew (2008) *Tescopoly: how one shop came out on top and why it matters*. London: Constable & Robinson.
71. Consumer Federation of America (2021) Surveillance advertising: how does the tracking work? https://consumerfed.org/consumer_info/factsheet-surveillance-advertising-how-tracking-works/
72. Schub, Justin (2019) Building a more private web. Google blog, 22 August. www.blog.google/products/chrome/building-a-more-private-web/
73. See for example this campaign to ban facial recognition: www.banfacialrecognition.com/stores/

2 HOW ADVERTISING INCREASES CONSUMPTION

1. See Fish 4.0 at: www.youtube.com/watc h?v=9PGam-zTKcY
2. Chen, A. (2015) The implicit link of luxury and self-interest: the influence of luxury objects on social motivation and cooperative behaviours. PhD thesis. University of Victoria, British Columbia, Canada.

3. Twenge, Jean and Tim Kasser (2013) Generational changes in materialism and work centrality, 1976–2007: associations with temporal changes in societal insecurity and materialistic role modelling. *Psychology Bulletin*, 39(7), 883–97. https://doi.org/10.1177/0146167213484586. Capella, M.L., C.R. Taylor and C. Webster (2008) The effect of cigarette advertising bans on consumption: a meta-analysis. *Journal of Advertising*, 37(2), 10. See also: www.who.int/tobacco/control/measures_art_13/en/ and www.who.int/tobacco/mpower/en/
4. Twenge and Kasser, Generational changes in materialism and work centrality.
5. Opree, S.J., M. Buijzen, E.A. van Reijmersdal and P.M. Valkenburg (2014) Children's advertising exposure, advertised product desire, and materialism: a longitudinal study. *Communications Research*, 41, 717–35.
6. Benmoyal-Bouzaglo, S. and G.P. Moschis (2010) Effects of family structure and socialization on materialism: a life course study in France. *Journal of Marketing Theory & Practice*, 18(1), 53–70.
7. Brand, J.E. and B.S. Greenberg (1994) Commercials in the classroom: the impact of Channel One advertising. *Journal of Advertising*, 34, 18–21. https://psycnet.apa.org/record/1994-31730-001
8. Jiang, R. and S.C. Chia (2009) The direct and indirect effects of advertising on materialism of college students in China. *Asian Journal of Communication*, 19(3), 319–36.
9. Nairn, A., J. Ormrod and P. Bottomley (2007) *Watching, wanting and wellbeing: exploring the links*. London: National Consumer Council.
10. Schor, Juliet (1993) *The overworked American: the unexpected decline of leisure*. New York: Basic Books.
11. Kasser, Tim (2020) *Advertising's role in climate and ecological degradation*. London: Badvertising. https://static1.squarespace.com/static/5ebdoo 80238e863d04911b51/t/5fbfcb1408845d09248d4e6e/160640489 1491/Advertising%E2%80%99s+role+in+climate+and+ecological+degradation.pdf
12. Ibid.
13. Ibid.
14. Kasser, Tim (2016) Materialistic values and goals. *Annual Review of Psychology*, 67(1), 489–514.
15. The Independent (1998) Catching a cold over advertisers' tactics. www.independent.co.uk/news/catching-a-cold-over-advertisers-tactics-1140043.html
16. Simms, Andrew and Tim Kasser (2020) Black Friday and the climate emergency. *The Ecologist*, 26 November.
17. Kasser (2020), *Advertising's role in climate and ecological degradation*.
18. Brack, J. and K. Cowling (1983) Advertising and labour supply: workweek and workyear in US manufacturing industries, 1919–76. *Kyklos*, 36(2), 285–303.

19. See for example: Hayden, Anders and John Shandra (2009) Hours of work and the ecological footprint: an exploratory analysis. *Local Environment*, 14, 575–600. www.researchgate.net/publication/249002192_Hours_of_work_and_the_ecological_footprint_An_exploratory_analysis
20. Brester, G.W. and T.C. Schroeder (1995) The impacts of brand and generic advertising on meat demand. *American Journal of Agricultural Economics*, 77, 969–79.
21. Cho, J.H., H.Y. Kim, T.K. Kim and B.S. Kim (2009) Impact of beef and pork generic advertising on Korean meat demand. *Korean Journal of Agricultural Management and Policy*, 36(3), 540–57.
22. Eshel, G., A. Shepon, T. Makov and R. Milo (2014) Land, irrigation water, greenhouse gas, and reactive nitrogen burdens of meat, eggs, and dairy production in the United States. *Proceedings of the National Academy of Sciences*, 111(33), 11996–2001. www.pnas.org/doi/abs/10.1073/pnas.1402183111?doi=10.1073%2Fpnas.1402183111#supplementary-materials
23. Reuters (2022) Global smoking rates fall for first time, but rise for kids, Africa – report. 18 May. www.reuters.com/business/healthcare-pharmaceuticals/global-smoking-rates-fall-first-time-rise-kids-africa-report-2022-05-18/
24. Moodie, C., J. Hoek, D. Hammond et al. (2022) Plain tobacco packaging: progress, challenges, learning and opportunities. *Tobacco Control*, 31, 263–71.
25. Cho, S.M., Y.M. Saw, N.N. Latt et al. (2020) Cross-sectional study on tobacco advertising, promotion and sponsorship (TAPS) and violations of tobacco sale regulations in Myanmar: do these factors affect current tobacco use among Myanmar high school students? *BMJ Open*, 10:e031933. doi: 10.1136/bmjopen-2019-031933
26. Cruz, T.G., R. McConnell, B.W. Low et al. (2019) Tobacco marketing and subsequent use of cigarettes, e-cigarettes, and hookah in adolescents. *Nicotine & Tobacco Research*, 21(7), 926–32.
27. Shmueli, D., J.J. Prochaska and S.A. Glantz (2010) Effect of smoking scenes in films on immediate smoking: a randomized controlled study. *American Journal of Preventive Medicine*, 38(4), April, 351–8. doi: 10.1016/j.amepre.2009.12.025. PMID: 20307802; PMCID: PMC2854161
28. Tynan, M.A., J.R. Polansky, D. Driscoll, C. Garcia and S.A. Glantz (2019) Tobacco use in top-grossing movies – United States, 2010–2018. *MMWR Morbidity and Mortality Weekly Report*, 68, 974–8. doi: http://dx.doi.org/10.15585/mmwr.mm6843a4
29. Sydney Morning Herald (2022) From the ashes: smoking's curious comeback on the silver screen. 18 March. www.smh.com.au/culture/tv-and-radio/from-the-ashes-smoking-s-curious-comeback-on-the-silver-screen-20220120-p59py5.html
30. WHO: Health promotion. www.who.int/teams/health-promotion/tobacco-control/implementing/measures
31. Kasser, *Advertising's role in climate and ecological degradation.*
32. See: Kasser, Tim et al. (2021) Advertising and demand for SUVs. London: New Weather Institute, Badvertising.

33. Davison, C. and B. Essen (2020) Eco-effectiveness: the missing measure in a climate crisis. Presentation available at https://ipa.co.uk/effworks/effworksglobal-2020/ecoeffectiveness-the-missing-measure-in-the-climate-crisis
34. Frick, V., E. Matthies, J. Thogersen and T. Santarius (2020) Do online environments promote sufficiency or overconsumption? Online advertisement and social media effects on clothing, digital devices, and air travel consumption. *Journal of Consumer Behavior*, 20(2), 288–308.
35. Kasser, *Advertising's role in climate and ecological degradation.*
36. Code of Ethics and Conduct – University College London. www.ucl.ac.uk/educational-psychology/resources/BPS_Code_of_Ethics_and_Conduct_2018.pdf
37. See The Ministry for the Climate Emergency: badverts.org

3 HOW WE BANNED TOBACCO ADVERTISING

1. Doll, Richard and Austin Bradford-Hill (1962) *Smoking and health*. London: Royal College of Physicians. www.rcplondon.ac.uk/projects/outputs/smoking-and-health-1962
2. British Medical Journal (1950) 30 September. www.ncbi.nlm.nih.gov/pmc/articles/PMC2038856/pdf/brmedj03566-0003.pdf
3. Simms, A. (2004) Would you buy a car that looked like this? *New Statesman*. www.newstatesman.com/node/161029
4. ASH (2020) Key dates in tobacco regulation: 1962–2020. London: ASH. https://ash.org.uk/wp-content/uploads/2020/04/Key-Dates.pdf
5. Quoted in Maisto, Stephen et al. (2007) *Drug use and abuse*, 5th edn. New York: Wadsworth, 144.
6. Null, Gary, Carolyn Dean, Martin Feldman, Debora Rasio and Dorothy Smith (2021) Death by Medicine 2010. Advocate Tanmoy Law Library, 24 May. https://advocatetanmoy.com/2021/05/24/death-by-medicine-gary-null-carolyn-dean-martin-feldman-debora-rasio-and-dorothy-smith/
7. Proctor, Robert N. (2012) The history of the discovery of the cigarette – lung cancer link: evidentiary traditions, corporate denial, global toll. *Tobacco Control*, 21(2), 20th Anniversary Issue (March), 87–91.
8. Ibid.
9. Gaetano, Phil (2018), The British Doctors' Survey 1951–2001. The Embryo Project Encyclopaedia, 30 January. https://embryo.asu.edu/pages/british-doctors-study-1951-2001.
10. The Times (1964) 24 January.
11. Hansard (1961) 18 October.
12. Hansard (1964) 2 June. https://hansard.parliament.uk/commons/1964-06-02/debates/32c4c46a-ebaa-4d55-ad4c-932e57e47dco/Clause4%E2%80%94(Tobacco)
13. National Center for Chronic Disease Prevention and Health Promotion (US) Office on Smoking and Health (2014). The health consequences of

smoking – 50 years of progress: a report of the Surgeon General. Atlanta, GA: US Centers for Disease Control and Prevention. 2, Fifty years of change 1964–2014. www.ncbi.nlm.nih.gov/books/NBK294310/

14. See for example: The Times (1964), 24 January.
15. Ibid.
16. Oreskes, Naomi and Erik Conway (2011) *Merchants of doubt*. London: Bloomsbury.
17. Lord Harris of High Cross (1994) Introduction to *Through the smokescreen of science*. London: FOREST.
18. ASH, Key dates in tobacco regulation.
19. Chapman, S. (1996) Cover essay: civil disobedience and tobacco control: the case of BUGA UP. *Tobacco Control*, 5(3), 179–85. www.jstor.org/stable/20207190
20. Conversation with Dr Moxham in July 2020.
21. Loughlin, Kelly (2005) Publicity as policy: the changing role of press and public relations at the BMA, 1940s–80s. *Making Health Policy*, 274–94. doi: https://doi.org/10.1163/9789004333109_013
22. ASH, Key dates in tobacco regulation.
23. Ibid.
24. BMA (1986), *Smoking out the barons: the campaign against the tobacco industry. A report of the British Medical Association*. Chichester, Wiley: BMA Public Affairs Division, 3.
25. Ibid., 5.
26. Ibid., 7.
27. Ibid..
28. Ibid., 33–5.
29. ASH, Key dates in tobacco regulation.
30. Ibid.
31. Ibid.
32. Herald and Times Archive (1993) Government faces onslaught against cigarette advertising. Glasgow, 23 January. www.heraldscotland.com/news/12625982.government-faces-onslaught-against-cigarette-advertising/
33. Burn, Chris (2019), How bruising miners' strike changed Kevin Barron from Scargillite to Labour moderniser. *Yorkshire Post*, 22 November. www.yorkshirepost.co.uk/news/politics/how-bruising-miners-strike-changed-kevin-barron-scargillite-labour-moderniser-1747020
34. Dorozynski, A. (1996) French biscuit makers force tobacco campaign ban. *British Medical Journal*, 313, 7.
35. High, Hugh (1998) *Advertising and smoking*. London: IEA.
36. Reuters (2022) Global smoking rates fall for first time, but rise for kids, Africa – report. 18 May. www.reuters.com/business/healthcare-pharmaceuticals/global-smoking-rates-fall-first-time-rise-kids-africa-report-2022-05-18/
37. US Surgeon General (1989) *Reducing the health consequences of smoking: 25 years of progress: a report of the Surgeon General*, United States Public Health Service, Office of the Surgeon General, Center for Chronic Disease

Prevention and Health Promotion, Office on Smoking and Health, Centers for Disease Control and Prevention, Series DHHS publication, no. CDC 89-8411. https://stacks.cdc.gov/view/cdc/13240
38. High (1998) *Advertising and smoking*.
39. US Surgeon General, Reducing the health consequences of smoking.
40. Anderson, R., S. Duckworth and C. Smee (1992) *Effect of tobacco advertising on tobacco consumption: a discussion document reviewing the evidence* (The Smee Report). London: Economics & Operational Research Division, Department of Health. https://archive.org/stream/op1279296-1001/op1279296-1001_djvu.txt
41. Ibid.
42. Bate, Roger (1998) A myth stubbed out. *Financial Times*, 20 April.
43. Burn, How bruising miners' strike changed Kevin Barron.
44. See Tobacco CEO's Statement to Congress 1994 news clip: Nicotine is not addictive. University of California, San Francisco. https://senate.ucsf.edu/tobacco-ceo-statement-to-congress
45. Guardian (2018) www.theguardian.com/environment/climate-consensus-97-per-cent/2018/sep/19/shell-and-exxons-secret-1980s-climate-change-warnings
46. Conversation with Lord Clement-Jones, July 2020.
47. Hansard (2001) 2 November. https://api.parliament.uk/historic-hansard/lords/2001/nov/02/tobacco-advertising-and-promotion-bill-hl
48. ASH (n.d.) Key dates in tobacco regulation 1962–2020. https://ash.org.uk/uploads/Key-Dates.pdf
49. Guardian (2015) Plain packaging to thank for Australia's decline in smoking, says Labor, 12 March.
50. DeMelle, Brendan and Kevin Grandia (2016) 'There is no doubt': Exxon knew CO2 pollution was a global threat by late 1970s. www.desmogblog.com/2016/04/26/there-no-doubt-exxon-knew-co2-pollution-was-global-threat-late-1970s
51. Department for Transport (2022) Official statistics, transport and environment statistics 2022.

4 SPORTS ADVERTISING AND SPONSORSHIP: THE GREAT POLLUTION OWN GOAL

1. BMJ Journals (2005) Tobacco in sport: an endless addiction? https://tobaccocontrol.bmj.com/content/14/1/1
2. Statista (2021) Market size of the global sports market from 2011 to 2018. www.statista.com/statistics/1087391/global-sports-market-size/
3. Simms, Andrew and Emilie Tricarico (2021) Sweat not oil: why sports should drop advertising and sponsorship from high-carbon polluters. London: New Weather Institute.

4. Skift (2013) Gulf airlines build awareness, one sports sponsorship at a time. 7 February. https://skift.com/2013/02/07/the-soft-power-of-gulf-airlines-manifested-in-their-sports-sponsorships
5. CEO Today (2021) Why do Audi, BMW and VW sponsor sports? www.ceotodaymagazine.com/2018/02/why-do-audi-bmw-vw-sponsorsports/
6. Reuters (2018) Big Oil spent 1 percent on green energy in 2018. 12 November.
7. New Weather Sweden (2022) Pie in the sky – how IATA's Fly Net Zero does nothing to save the climate. www.newweather.se/blogg/pie-in-the-sky-94tce-btcwf
8. CBS (2015) Not just VW: a long history of cheating car companies. 29 September.
9. IATA (2020) Air passenger market analysis. February. www.iata.org/en/iata-repository/publications/economic-reports/airpassenger-monthly-analysis---feb-2020
10. Eurosport (2022) www.eurosport.com/cycling/british-cycling-chief-executive-brian-facer-leaves-after-shell-deal-backlash-and-emily-bridges-contr_sto9209074/story.shtml
11. C&EN (2019) C&EN's global top 50 chemical companies of 2018. 29 July. https://cen.acs.org/business/finance/CENs-Global-Top-50-chemical/97/i30
12. Cyclist (2020) Ineos Grenadiers' budget has hit €50 million for the first time. 8 October. www.cyclist.co.uk/news/8798/ineos-grenadiers-budget-has-hit-50- million-for-the-first-time. Cyclist (2019) Team Ineos owner Jim Ratcliffe's €100m purchase of OGC Nice imminent. 18 June. www.cyclist.co.uk/news/6257/team-ineos-owner-jim-ratcliffes-100mpurchase-of-ogc-nice-imminent
13. DW.com (2019) Will Mr Nice transform French football? 16 September. www.dw.com/en/will-mr-nice-transform-french-football/a-50361836
14. C&EN, C&EN's global top 50 chemical companies of 2018.
15. Guardian (2020) Sir Jim Ratcliffe, UK's richest person, moves to tax-free Monaco. 25 September. www.theguardian.com/business/2020/sep/25/sir-jim-ratcliffe-uks-richest -person-moves-to-tax-free-monaco-brexit-ineos-domicile
16. Energy Live News (2022) INEOS offers to drill fracking test well. 9 September. www.energylivenews.com/2022/09/09/ineos-offers-to-drill-fracking-test-well/
17. Guardian (2022) Rishi Sunak will keep ban on fracking in UK. No 10 confirms. 26 October. www.theguardian.com/environment/2022/oct/26/rishi-sunak-ban-on-fracking-uk-no-10
18. Cyclingtips (2019) What does Ineos have to gain by sponsoring a cycling team? https://cyclingtips.com/2019/05/what-does-ineos-have-to-gain-bysponsoring-a-cycling-team/. Guardian (2019) Yorkshire village faces petrochemical giant in anti-fracking fight. 11 June. www.theguardian.com/environment/2019/jun/11/yorkshire-villagepetrochemical-ineos-fracking. See also: https://drillordrop.com/2020/10/23/ineos-cuts-63m-from-value-

of-uk-shaleassets-because-of-fracking-moratorium/. Also: Frack Off (2019) Ineos. https://frack-off.org.uk/companies/ineos/

19. BBC (2020) Sir Jim Ratcliffe confirms new vehicle to be made in France. 8 December. www.bbc.co.uk/news/amp/business-55236852. Ineos Grenadier (2020) Ineos automotive confirms acquisition of hambach production site from Mercedes-Benz. 8 December. https://ineosgrenadier.com/news/ineos-automotive-confirms-acquisition-ofhambach-production-site-from-mercedes-benz.
20. Friends of the Earth UK (2016) Briefing Ineos, September. https://friendsoftheearth.uk/sites/default/files/downloads/ineos-briefing101850.pdf
21. RTS (2018) Ineos, propriétaire du Lausanne-Sport et 'ennemi numéro 1' en Angleterre. 5 March. rts.ch/info/monde/9379910-ineos-proprietaire-du-lausannesportet-ennemi-numero-1-en-angleterre.html
22. Le Temps (2017) Racheté par Ineos, le Lausanne-Sport se rêve en grand. 13 November. www.letemps.ch/sport/rachete-ineos-lausannesport-se-reve-grand
23. Ibid.
24. Port of Antwerp (2019) INEOS announces mega-investment in port of Antwerp. www.portofantwerp.com/en/news/ineos-announces-mega-investment-port-antwerp
25. FairFin (2020) How our government and banks are trying to fill the bottomless pit of Ineos. December. www.fairfin.be/sites/default/files/2021-02/the%20bottomless%20pit%20of% 20Ineos%20ENG_v2.pdf. Cycling News (2020) Environmentalists damage Ineos Grenadiers vehicles in East Flanders. 21 December. www.cyclingnews.com/news/environmentalists-damage-ineosgrenadiers-vehicles-in-east-flanders/
26. Transport & Environment (2021) Road vehicles and air quality. www.transportenvironment.org/what-we-do/air-quality-and-transport/road-vehicles-and-air-quality
27. Ibid.
28. EPHA (2018) Health impacts and costs of diesel emissions in the EU. November. https://epha.org/wp-content/uploads/2018/11/embargoed-until-27-november-00 -01-am-cet-time-ce-delft-4r30-health-impacts-costs-diesel-emissions-eu-def. pdf
29. Vohra, K. et al. (2021) Global mortality from outdoor fine particle pollution generated by fossil fuel combustion: results from GEOS-Chem. www.sciencedirect.com/science/article/abs/pii/S0013935121000487 See also: Harvard (2021) Fossil fuel air pollution responsible for more than 8 million people worldwide in 2018. 9 February. www.seas.harvard.edu/news/2021/02/deaths-fossil-fuel-emissions-higher -previously-thought. Reuters (2021) Fossil fuel pollution causes one in five premature deaths globally: study. 9 February. www.reuters.com/article/us-health-pollution-fossil/fossil-fuel-pollutioncauses-one-in-five-premature-deaths-globally-study-idUSKBN2A90UB

30. New Scientist (2020) Air pollution linked to greater risk of dying from covid-19 in the US. 4 November. www.newscientist.com/article/2258774-air-pollution-linked-to-greaterrisk-of-dying-from-covid-19-in-the-us/
31. New Weather Institute (2023) Dangerous driving – why sport should drop sponsorship from major polluters: the cases of Toyota and BMW. London.
32. Misoyannis, A. (2023) Toyota remained world's biggest car maker in 2022. Drive, 31 January. www.drive.com.au/news/toyota-remains-worlds-biggest-car-maker-in-2022/
33. Toyota Global (n.d.) *Affiliates*. www.toyota-global.com/company/history_of_toyota/75years/data/automotive_business/production/production/japan/general_status/other.html
34. Car Sales Statistics (2022) 2021 (full year) global: Toyota worldwide car sales, production, and exports. www.best-selling-cars.com/brands/2021-full-year-global-toyota-worldwide-car-sales-production-and-exports
35. Ibid.
36. Ibid.
37. Ibid.
38. Guardian (2019) AFL in $18.5m-a-year sponsorship deal with Toyota, reportedly largest ever in Australia. 15 March. www.theguardian.com/sport/2019/mar/15/afl-in-185m-a-yearsponsorship-deal-with...
39. USA Today (2021) Toyota drives onto Olympic stage in record sponsorship deal. https://eu.usatoday.com/story/sports/olympics/2015/03/13/toyota-drives-ontoolympic-stage-in-record-sponsorship-deal/70258812/
40. Ibid. IEG Sponsorship Report (2016) The most active sponsors in the auto category. 29 August. www.sponsorship.com/iegsr/2016/08/29/The-Most-Active-Sponsors-InThe-Auto-Category--Who.aspx
41. University of Waterloo (2014) Can the Winter Olympics survive climate change? 22 January. https://uwaterloo.ca/news/sites/ca.news/files/uploads/files/oly_winter_games_ warmer_world_2014.pdf
42. Reuters (2021) Salla, coldest town in Finland, to bid for 2032 Summer Games. 26 January. www.reuters.com/article/us-olympics-bid-salla/salla-coldest-town-infinland-to-bid-for-2032-summer-games-idUSKBN29V2I8
43. The Climate Institute (2014) Sport & climate impacts: how much heat can sport handle? www.connect4climate.org/sites/default/files/files/publications/Sport_ and_climate_report.pdf
44. Ibid.
45. Ibid.
46. Ibid.
47. NBC News (2015) More than 30 hospitalized during sweltering Los Angeles marathon. 16 March. www.nbcnews.com/news/us-news/more-30-hospitalized-duringsweltering-los-angeles-marathon-n324021
48. Grundstein, A.J. et al. (2012) A retrospective analysis of American football hyperthermia deaths in the United States. https://pubmed.ncbi.nlm.nih.gov/21161288/

49. Rapid Transition Alliance (2020) *Playing against the clock: global sport, the climate emergency and the case for rapid change*. 20 June. www.rapidtransition.org/resources/playing-against-the-clock/
50. Ibid.
51. Ibid.
52. Ibid.
53. Ibid.
54. Marshall, D.W. and G. Cook (2015) The corporate (sports) sponsor. www.tandfonline.com/doi/abs/10.1080/02650487.1992.11104507. Also: Abratt, R. et al. (2015) Corporate objectives in sports sponsorship. 2 March. www.tandfonline.com/doi/abs/10.1080/02650487.1987.11107030 And: Copeland, R. et al. (1996) Understanding the sport sponsorship process from a corporate perspective. https://journals.humankinetics.com/view/journals/jsm/10/1/article-p32.xml
55. Abratt et al., Corporate objectives in sports sponsorship. Also: Copeland, R. et al., Understanding the sport sponsorship process from a corporate perspective.
56. Pope, N., K. LL. and Voges K.E. (2000) The impact of sport sponsorship activities, corporate image and prior use on consumer purchase intention. www.researchgate.net/profile/Kevin_Voges/publication/29454754_The_impact_of_sport_sponsorship_activities_Corporate_image_and_prior_use_on _a_consumer_purchase_Intent/links/5432f1540cf22395f29e01c7/The-impact-ofsport-sponsorship-activities-Corporate-image-and-prior-use-on-a-consumer-p urchase-Intent.pdf
57. IEG Sponsorship Report (2010) Inside GM's new sponsorship strategy. 7 May. www.sponsorship.com/iegsr/2010/05/10/Inside-GM-s-New-Sponsorship-St rategy.aspx
58. Mason, K. (2005) How corporate sport sponsorship impacts consumer behavior. www.researchgate.net/publication/237227040_How_Corporate_Sport_ Sponsorship_Impacts_Consumer_Behavior
59. Campaign (2018) Toyota signs eight-year Team GB sponsorship deal. 5 February. www.campaignlive.co.uk/article/toyota-signs-eight-year-team-gbsponsorship-deal/1456239
60. The Drum (2020) Toyota signs eight-year Team GB sponsorship deal. 6 January. www.thedrum.com/news/2020/01/06/toyota-and-bp-sponsor-channelT4-s-tokyo-2020-paralympic-coverage
61. The Drum (2020) Emirates agrees five-year Olympique Lyonnais sponsorship deal. 7 February. www.thedrum.com/news/2020/02/07/emirates-agrees-five-yearolympique-lyonnais-sponsorship-deal
62. Guardian (2015) Cricket World Cup: India v Pakistan watched by a billion people – in pictures. 4 February. www.theguardian.com/sport/gallery/2015/feb/15/cricket-world-cupindia-v-pakistan-watched-by-a-billion-people-in-pictures
63. Advertising Association (2020) Ad Net Zero: COP26 briefing with Strategy Director, Charles Ogilvie. 16 December. https://adassoc.org.uk/

events/ad-net-zero-cop26-briefing-with-strategydirector-charles-ogilvie/. Sponsorship Intelligence (2012) London 2012 Olympic Games Global Broadcast Report. December. https://stillmed.olympic.org/Documents/IOC_Marketing/Broadcasting/London_ 2012_Global_%20Broadcast_Report.pdf

64. Guardian (2018) Amnesty turns the heat up on 'sportswashing' Manchester City owners. 10 November. www.theguardian.com/football/2018/nov/10/manchester-city-amnestyinternational-football-leaks
65. These Football Times (2015) Gazprom's colossal football empire. 15 January. https://thesefootballtimes.co/2015/01/15/the-gazprom-empire/
66. Ibid.
67. Sports Pro (2010) Red Star Belgrade sign multi-million dollar deal with Gazprom. 16 July. www.sportspromedia.com/news/red_star_belgrade_sign_multi-million_ dollar_deal_with_gazprom
68. These Football Times, Gazprom's colossal football empire.
69. Marketscreener (2017) Gazprom Neft': film showing Zenit players in a match on the Prirazlomnaya offshore platform to be shown on NTV. 6 July. www.marketscreener.com/quote/stock/GAZPROM-NEFT-6494696/news/Gazprom-Neft-Film-showing-Zenit-players-in-a-match-on-the-Prirazlomnaya-offshore-platform-to-be-s-24556600/
70. Vox (2020) Why this Russian gas company sponsors soccer teams. 31 January. www.vox.com/videos/2020/1/31/21117233/gazprom-russia-soccer-sponsor
71. Ibid.
72. Ibid.
73. Artnet (2020) Museums' rejection of fossil fuel sponsorship is more than just symbolic. Here's why it can have real-world impact. 28 February. https://news.artnet.com/opinion/culture-unstained-museum-sponsorship-oped-1789302
74. Smith, Kevin (2023) 2 March. https://twitter.com/kevinjgsmith/status/1631388813032300615?s=20
75. Champions for Earth (2020) Green recovery letter. 23 September. https://championsforearth.com/green-recovery-letter/
76. Ecoathletes (2020) Champions for earth becomes newest ecoathletes organizational supporter. 31 August. https://ecoathletes.org/2020/08/champions-for-earth-becomes-newestecoathletes-organizational-supporter/
77. Sky Sports (2023) Pro skiers demand competition changes, fearing winter sports becoming 'unjustifiable' to public. 14 February. https://news.sky.com/story/pro-skiers-demand-competition-changes-fearing-winter-sports-becoming-unjustifiable-to-public-12810714
78. UNFCCC, Sports for Climate Action Framework. https://unfccc.int/sites/default/files/resource/Sports_for_Climate_Action_
79. UN Sports for Climate Action Framework – Principle 4: Promote sustainable and responsible consumption. The objective of this principle encourages sports organisations and sports events organisers to adopt sustainable procurement policies to motivate providers to develop cleaner options. Communication campaigns towards fans and other stakeholders

could be built, to promote the use of greener, sustainable options. This also applies to giving preference to sustainable means of transport, being one of the major sources of greenhouse emissions in sports, thereby supporting global transition to low-carbon transport.

80. France 24 (2019) Total pull sponsorship plug on 2024 Olympics over 'eco-Games'. 5 June. www.france24.com/en/20190605-total-pull-sponsorship-plug-2024- olympics-over-eco-games
81. Boehm, S. and D. Siddharta (2009) Upsetting the offset: the political economy of carbon markets. http://repository.essex.ac.uk/7271/
82. Climate Home News (2018) FIFA accused of greenwashing in World Cup carbon offset scheme. 11 June. www.climatechangenews.com/2018/06/11/fifa-accused-greenwashingworld-cup-carbon-offset-scheme
83. Guardian (2023) Revealed: more than 90% of rainforest carbon offsets by biggest certifier are worthless, analysis shows. 18 January. www.theguardian.com/environment/2023/jan/18/revealed-forest-carbon-offsets-biggest-provider-worthless-verra-aoe
84. Badvertising (2022) Caught offside with offsets? Why offsetting won't solve sports' climate problem. November. www.badverts.org/latest/briefing-why-offsets-wont-solve-sports-climate-problem
85. Euractive (2023) S&Ds lead major improvements in EU law empowering consumers for the green transition. 11 May.
86. Rapid Transition Alliance, *Playing against the clock.*
87. Ibid.
88. Rapid Transition Alliance (2020) An open goal for transition – will global sport follow the lead of a small English football club? 27 August. www.rapidtransition.org/stories/an-open-goal-for-transition-will-globalsport-follow-the-lead-of-a-small-english-football-club
89. Abrahamsson, Mats, Andrew Simms, Gunnar Lind and Anna Jonsson (2023) *The snow thieves: how high carbon sponsors are melting winter sports.* London: Badvertising. https://static1.squarespace.com/static/5ebd0080238e863d04911b51/t/63f8b3fdbb19fb22ce7b6394/1677243451215/The+Snow+Thieves.pdf
90. Guardian (2023) 'Big irony' as winter sports sponsored by climate polluters, report finds. 27 February. www.theguardian.com/environment/2023/feb/27/big-irony-as-winter-sports-sponsored-by-climate-polluters-report-finds

5 HOW BIG CAR PERSUADED US TO BUY BIG CARS

1. Financial Times (2023) SUVs hit record share of new car sales in EU. 19 March. www.ft.com/content/151cb429-d024-4d5c-9edf-5b4a2b104a66
2. Transport & Environment (2018) Six EU governments finally face legal action over air pollution. 7 June. www.transportenvironment.org/discover/six-eu-governments-finally-face-legal-action-over-air-pollution/
3. InsideRadio (2018) Ford Ad spend predicted company's shift away from cars. 12 November. www.insideradio.com/free/ford-ad-spend-predicted-company-

s-shift-away-from-cars/article_69c0c832-e65a-11e8-a609-5b56473df19e.html

4. Bradsher, Keith (2002) *High and mighty: The dangerous rise of the SUV*. New York: Public Affairs.
5. IEA (2019) Growing preference for SUVs challenges emissions reductions in passenger car market. www.iea.org/commentaries/growing-preference-for-suvs-challenges-emissions-reductions-in-passenger-car-market
6. Transport & Environment (2018) CO2 emissions from cars: the facts. www.transportenvironment.org/wp-content/uploads/2021/07/2018_04_CO2_emissions_cars_The_facts_report_final_0_0.pdf
7. New Weather institute/Possible (2020) *Upselling Smoke: the case to end advertising of the largest, most polluting new cars*. www.badverts.org/s/Upselling-Smoke-FINAL-23-07-20.pdf
8. www.badverts.org/latest/mindgames-on-wheels-our-latest-report
9. Boyle, David (2017) *Edison*. London: Sharpe Books & the Real Press.
10. Vaknin, Judy (2008), *Driving it home; 100 years of car advertising*. London: Middlesex University Press, 51.
11. Bradsher, *High and mighty*.
12. Vaknin, *Driving it home*, 112.
13. Ibid., 120.
14. Ibid., 108.
15. Bradsher, *High and mighty*.
16. Ibid., 101.
17. Ibid., 98–9.
18. Gunster, Shane (2004) 'You belong outside': advertising, nature, and the Suv. *Ethics and the Environment*, 9(2), Fall/Winter, 4–32.
19. Financial Times (2023) SUVs hit record share of new car sales in EU. 19 March. www.ft.com/content/151cb429-d024-4d5c-9edf-5b4a2b104a66
20. Resumé (2020) Volvo Cars lanserar global kampanj – med Fares Fares. 9 September. www.resume.se/marknadsforing/kampanj/volvo-cars-lanserar-global-kam panj-med-fares-fares/
21. Anderson, Michael and Maximilian Auffhamme (2011) In pounds that kill: the external costs of vehicle weight. NBER Working Paper No. 17170. www.nber.org/papers/w17170
22. Insurance Institute for Highway Safety (IIHS) / The Highway Loss Data Institute (HLDI) (2022) SUVs, other large vehicles often hit pedestrians while turning. 17 March.
23. Ivarsson, Bo, Basem Henary, Jeff Crandall and Douglas Longhitano (2005). Significance of adult pedestrian torso injury. *Annual Proceedings Association for the Advancement of Automotive Medicine*, 49, 263–77.
24. Edwards, Mickey and Daniel Leonard (2022) Effects of large vehicles on pedestrian and pedalcyclist injury severity. *Journal of Safety Research*, 82, 275–82. https://doi.org/10.1016/j.jsr.2022.06.005

25. IIHS-HLDI (2023) High point of impact makes SUV crashes more dangersous for cyclists.13 April. www.iihs.org/news/detail/higher-point-of-impact-makes-suv-crashes-more-dangerous-for-cyclists
26. Dreiser, Theodore (1916) *A hoosier holiday*. New York: John Lane, 98.
27. Simms, Andrew (2021) Car free stories: the irresistible rise of people-friendly, clean air cities. Possible, September. www.wearepossible.org/our-reports-1/car-free-stories-the-irresistible-rise-of-people-friendly-clean-air-cities
28. See remarks by R.A. Kidd, former county surveyor of Nottinghamshire in Smith, Allan (1984) *A history of the County Surveyors Society*.
29. Shalite, Ruth (1999) The return of the hidden persuaders. Salon, 27 September. www.salon.com/1999/09/27/persuaders/
30. Hummer 4x4 Off Road (2011) www.hummer4x4offroad.com/forum/showthread.php/487-HUMMER-Decal-I deas
31. Vaknin, *Driving it home*, 140.
32. Gentlemen's Quarterly, August 2000. Quoted in Gunster, 'You belong outside'.
33. Jayara, Abhirami (2019) (Mis)Appropriations of nature: political and positional aspects of the 'nature' in select advertisements. *Navajyoti, International Journal of Multi-Disciplinary Research*, 4(1), August.
34. Farrell, James (2010) *The nature of college: how a new understanding of campus life can change the world*. New York, Milkweed, 133.
35. Gunster, 'You belong outside'.
36. Los Angeles Magazine (1999) January.
37. Ebay (2021) 1998 Dodge Durango tread lightly and carry a big V 8-original print ad. www.ebay.com/itm/1998-Dodge-Durango-Tread-Lightly-And-Carry-A-BigV-8-Original-Print-Ad-8-5-x-11-/184062557068
38. Adweek (2001) Honda campaign touts power, reliability. www.adweek.com/brand-marketing/honda-campaign-touts-power-reliabili ty-52880/
39. https://crashstats.nhtsa.dot.gov/Api/Public/ViewPublication/813428
40. Bradsher, Keith (2000) *New York Times*, 18 July.
41. Bradsher, *High and mighty*, 67.
42. Quoted in Adams, John (1987) How pedestrians can make roads safer, *Town & Country Planning*, April.
43. Ibid.
44. Gunster, 'You belong outside'.
45. Ibid.
46. Rollins, William (2006), Reflections on a spare tire: SUVs and postmodern environmental consciousness. *Environmental History*, 11, October.
47. Ibid.
48. D&AD (2018) Clowns. www.dandad.org/awards/professional/2018/film-advertising/26547/clowns/
49. Gunster, 'You belong outside'.
50. Ibid.

51. Nolt, Leonard (2008) Letters from 1998. 5 March. http://leonardnolt.blogspot.com/2008/03/letters-from-1998.html
52. Bradsher, *High and mighty*, ix.
53. Ibid., 169–70.
54. Ibid., 163–4.
55. Ibid., 141.
56. www.bloomberg.com/graphics/2019-ford-explorer-owners-say-suvs-making-them-sick/
57. Gunster, 'You belong outside'.
58. Guardian (2002) 'What would Jesus drive?' Gas-guzzling Americans are asked. 14 November. www.theguardian.com/world/2002/nov/14/usa.oliverburkeman
59. Holder, Jim (2020) *Autocar*, 12 August.
60. Eunomia (2018) Investigating options for reducing releases in the aquatic environment of microplastics emitted by products, for the European Commission. www.eunomia.co.uk/reports-tools/investigating-options-for-reducing-releases-in-the-aquatic-environment-of-microplastics-emitted-by-products/
61. Friends of the Earth (2018) Car tyres responsible for thousands of tonnes of UK plastic pollution. 22 November.
62. Reuters (2020) Tyre industry pushes back against evidence of plastic pollution. 7 September. www.euractiv.com/section/transport/news/tyre-industry-pushes-back-against-evidence-of-plastic-pollution/
63. Rollins, Reflections on a spare tire.
64. AdNews (2017) How SUVs are driving auto advertising spend. 26 September. www.adnews.com.au/news/how-suvs-are-driving-auto-advertising-spend
65. Vilchez, J.J., R. Pasqualino and Y. Hernandez (2023) The new electric SUV market under battery supply constraints: might they increase CO_2 emissions? *Journal of Cleaner Production*. https://doi.org/10.1016/j.jclepro.2022.135294
66. Dearman, C. et al. (2023) Sports utility vehicles: a public health model of their climate and air pollution impacts in the United Kingdom, *International Journal of Environ. Research and Public Health*, 20(11), 6043. https://doi.org/10.3390/ijerph20116043
67. Badvertising (2022) Calls for UK public transport authorities to take common-sense measure to remove ads for polluting travel. 5 December. www.badverts.org/latest/calls-for-uk-public-transport-authorities-to-take-common-sense-measure-to-remove-ads-for-polluting-travel
68. The CAP Code 30:7. The Committee of Advertising Practice (CAP) offers guidance on the interpretation of the UK Code of Non-broadcast Advertising and Direct & Promotional Marketing (the CAP Code) in relation to non-broadcast marketing communications. The Broadcast Committee of Advertising Practice (BCAP) offers guidance on the interpretation of the UK Code of Broadcast Advertising (the BCAP Code) in relation to broadcast advertisements.

6 HOW AIRLINES TOOK US FOR A RIDE

1. Aviation history: the desire to fly in the mythology. https://applications.icao.int/postalhistory/aviation_history_the_mythology.htm
2. Get information: we have to talk about flying! (2020) Stay Grounded. https://stay-grounded.org/get-information/#impact
3. This excludes the issue of sexual reproduction, which we regard as a category error in the consideration of personal mitigation efforts. See: Wynnes, Seth and Kimberly A. Nicholas (2017) The climate mitigation gap: education and government recommendations miss the most effective individual actions. www.researchgate.net/publication/318353145_The_climate_mitigation_gap_Education_and_government_recommendations_miss_the_most_effective_individual_actions
4. The tortoise and the hare? The race to decarbonize aviation, Blog. https://theicct.org/global-aviation-race-jun22/
5. ICAO (2019) Annual report: the world of air transport in 2019. www.icao.int/annual-report-2019/Pages/the-world-of-air-transport-in-2019.aspx#:~:text=According%20to%20ICAO's%20preliminary%20compilation,a%201.7%20per%20cent%20increase
6. Possible (2022) Missed targets: a brief history of aviation climate charges, 9 May. www.wearepossible.org/our-reports-1/missed-target-a-brief-history-of-aviation-climate-targets
7. Fantuzzi, A. et al. (2023) Low-carbon fuels for aviation. Imperial College London Briefing Paper No, 9. March. https://spiral.imperial.ac.uk/bitstream/10044/1/101834/11/IMSE_Low_carbon_aviation_fuels_briefing_paper_2023.pdf
8. Forecast International (2023) Airbus and Boeing report January 2023 commercial aircraft orders and deliveries, 16 February. https://dsm.forecastinternational.com/wordpress/2023/02/16/airbus-and-boeing-report-january-2023-commercial-aircraft-orders-and-deliveries/
9. IATA (2022) Incentives needed to increase SAF production. Press release, 21 June. www.iata.org/en/pressroom/2022-releases/2022-06-21-02/
10. The Royal Society (2023) Net zero aviation fuels: resource requirements and environmental impacts. Policy briefing, February. https://royalsociety.org/-/media/policy/projects/net-zero-aviation/net-zero-aviation-fuels-policy-briefing.pdf
11. O'Malley, Jane, Nikita Pavlenko and Stephanie Searle (2021) Estimating sustainable aviation fuel feedstock availability to meet growing European Union demand. International Council on Clean Transportation Working Paper 2021-13, March. https://theicct.org/sites/default/files/publications/Sustainable-aviation-fuel-feedstock-eu-mar2021.pdf
12. Becken, Susan, Brendan Mackey and David S. Lee (2023) Implications of preferential access to land and clean air for sustainable aviation fuels. *Science of The Total Environment*, 886, 15 August, 163883. www.sciencedirect.com/science/article/pii/S0048969723025044

13. Carbon Change Committee (2020) The sixth generation carbon budget: electricity generation. www.theccc.org.uk/wp-content/uploads/2020/12/Sector-summary-Electricity-generation.pdf
14. Zhou, Yuanrong, Stephanie Searle and Nikita Pavlenko (2022) Current and future cost of e-kerosene in the United States and Europe. International Council on Clean Transportation Working Paper 2022-14, March. https://theicct.org/wp-content/uploads/2022/02/fuels-us-europe-current-future-cost-ekerosene-us-europe-mar22.pdf
15. Financial Times (2022) Aviation calls on UK government to subsidise 'Jet Zero' push, 6 November. www.ft.com/content/9a3ed7af-9637-4c03-bbc9-f1d8dcefe2c7
16. Possible (2022) Aviation demands management's role in deep carbonisation pathways, 26 April. www.wearepossible.org/our-reports-1/aviation-demand-managements-role-in-deep-decarbonisation-pathways
17. Policy implementation timeline (2022) California Copyright Conference. www.theccc.org.uk/wp-content/uploads/2022/06/Policy-implementation-timeline-Aviation.pdf
18. IEA (2021) Net zero by 2015: a roadmap for the global energy sector, October. https://iea.blob.core.windows.net/assets/deebef5d-0c34-4539-9d0c-10b13d840027/NetZeroby2050-ARoadmapfortheGlobalEnergySector_CORR.pdf
19. Allwood et al. (2019) *Absolute zero*. University of Cambridge. www.repository.cam.ac.uk/items/33aaf353-b7de-45b0-9c40-5f62975b2127
20. Possible (2021) Elite status: how a small minority around the world take an unfair share of flights.
21. Gössling, Stefan and Andreas Humpe (2020) The global scale, distribution and growth of aviation: implications for climate change. *Global Environmental Change*. www.sciencedirect.com/science/article/pii/S0959378020307779#:~:text=Finds%20that%201%25%20of%20world,CO2%20from%20commercial%20aviation.&text=Suggests%20that%20emissions%20from%20private,t%20CO2%20per%20year
22. Office for National Statistics (2021) Travel trends (2021). www.ons.gov.uk/peoplepopulationandcommunity/leisureandtourism/articles/traveltrends/2021
23. Gössling, Stefan, Paul Hanna, James Higham, Scott Cohen and Debbie Hopkins (2019) Can we fly less? Evaluating the 'necessity' of air travel. *Journal of Air Transport Management*, 81, October, 101722. www.sciencedirect.com/science/article/abs/pii/S0969699719303229
24. Tim Kasser et al. (2020) *Advertising's role in climate and ecological degradation – what does the scientific research have to say?* London: New Weather Institue. https://static1.squarespace.com/static/5ebd0080238e863d04911b51/t/5fbfcb1408845d09248d4e6e/1606404891491/Advertising%E2%80%99s+role+in+climate+and+ecological+degradation.pdf

25. New Weather Institute/GP Nordics (2022) Advertising climate chaos. https://www.greenpeace.org/static/planet4-sweden-stateless/2022/02/d423c8a6-advertising-climate-chaos-report.pdf
26. Urios, Jesus and Thorfinn Stainforth (2022) Can polluter pays principles in the avaiation sector be progressive? Institute for European Environmental Policy, 22 November. https://ieep.eu/publications/can-polluter-pays-principles-in-the-aviation-sector-be-progressive/
27. SMMT 2022 Automotive Sustainability Report (2022) Average vehicle age. www.smmt.co.uk/industry-topics/sustainability/average-vehicle-age/
28. Quick, Miriam (2019) Domestic air travel has dipped as climate-conscious Swedes opt for train. But will 'flight shame' become the norm – and what might it mean for business travel. BBC, 22 July. www.bbc.com/worklife/article/20190718-flygskam
29. The Clean Development Mechanism. United Nations Climate Change. https://unfccc.int/process-and-meetings/the-kyoto-protocol/mechanisms-under-the-kyoto-protocol/the-clean-development-mechanism
30. www.reuters.com/markets/carbon/voluntary-carbon-markets-set-become-least-five-times-bigger-by-2030-shell-2023-01-22 19/
31. Simma, Andrew and Freddie Daley (2022) Offsets caught offside. *Ecologist*, 13 December. https://theecologist.org/2022/dec/13/offsets-caught-offside#:~:text=One%20study%20for%20the%20EU,high%20likelihood%20of%20doing%20so
32. Greenfield, Patrick (2023) Revealed: more than 90% of rainforest carbon offsets by biggest certifier are worthless, analysis shows. *Guardian*, 18 January. www.theguardian.com/environment/2023/jan/18/revealed-forest-carbon-offsets-biggest-provider-worthless-verra-aoe
33. ICAO (2023) Carbon Offsetting and Reduction Scheme (CORSIA), June. www.icao.int/environmental-protection/CORSIA/Pages/default.aspx
34. National Air and Space Museum, Smithsonian. The evolution of the commercial flying experience: 1914–today. https://airandspace.si.edu/explore/stories/evolution-commercial-flying-experience#1927
35. F 2206 American Airlines Commercials from the 1960s, San Diego Air and Space Museum Archives.
36. Lovegrove, Keith (2011) *Airline: style at 30,000 feet*, 2nd edn. London: Laurence King.
37. Ibid., 32.
38. Miller, Meg (2015) Why airline branding used to be so much better than it is today. brandnewmag,com, 7 September. www.brandknewmag.com/why-airline-branding-used-to-be-so-much-better-than-it-is-today/
39. Cozens, Claire (2003) Easyjet adds escape censure. *Guardian*, 30 July, www.theguardian.com/media/2003/jul/30/advertising.travelnews
40. *Meeting the UK aviation target – options for reducing emissions to 2050*. Climate Change Committee, 2009.
41. The 2020 bailouts left airlines, the economy, and the federal budget in worse shape than before. Mercatus Centre, 2022. www.mercatus.org/research/

policy-briefs/2020-bailouts-left-airlines-economy-and-federal-budget-worse-shape
42. Thenextweb.com (2021) United dragged by Twitter over greenwashing with its '100% sustainable fuel' flight. 3 December. https://thenextweb.com/news/twitter-vs-united-over-greenwashing-with-100-percent-sustainable-fuel-flight
43. United.com (n.d.) Eco-Skies Alliance. www.united.com/ual/en/us/fly/company/global-citizenship/environment/ecoskies-alliance.html
44. VCCP London (2022) Next Gen EasyJet. 25 March. www.vccp.com/work/easyjet/nextgen-easyjet
45. Badvertising (2022) EasyJet's Greenwash adverts under scrutiny. 14 April. https://www.badverts.org/latest/easyjets-greenwash-adverts-under-scrutiny
46. Banking on Climate Chaos (2023) Fossil fuel finance report 2023. www.bankingonclimatechaos.org/
47. Badvertising Sweden (2022) Airline Greenwash grounded by Swedish ad regulator.
48. Euronews (2023) KLM axes 'misleading' ads but won't stop promoting sustainability initiatives. April.
49. Politico (2023) European Parliament takes aim at guilt-free flights. May.
50. Guardian (2023) Lufthansa's 'green' adverts banned in UK for misleading consumers. 1 March.
51. The Times (2023) There's no such thing as 'green' flying, says advertising watchdog. 12 April. www.thetimes.co.uk/article/theres-no-such-thing-as-green-flying-says-advertising-watchdog-sz6tjhmqf
52. Guardian (2023) Adverts claiming products are carbon neutral by using offsetting face UK ban. 15 May. www.theguardian.com/environment/2023/may/15/uk-advertising-watchdog-to-crack-down-on-carbon-offsetting-claims-aoe
53. Adfree Cities (2021) Airline misleads Euros football fans with pitchside greenwash. 19 July. https://adfreecities.org.uk/2021/07/airline-misleads-euros-football-fans-with-pitchside-greenwash/

7 WHY SELF-REGULATION ISN'T WORKING

1. CMA (2022) Compliance principles for social media platforms. 3 November. www.gov.uk/government/publications/compliance-principles-for-social-media-platforms
2. ASA, Advertising Codes. www.asa.org.uk/type/non_broadcast/code_section/01.html
3. Guardian (2022) Watchdog bans HSBC climate ads in fresh blow to bank's green credentials. 19 October. www.theguardian.com/business/2022/oct/19/watchdog-bans-hsbc-ads-green-cop26-climate-crisis
4. The CAP Code 30:7. The Committee of Advertising Practice (CAP) offers guidance on the interpretation of the UK Code of Non-broadcast Advertising and Direct & Promotional Marketing (the CAP Code) in

relation to non-broadcast marketing communications. The Broadcast Committee of Advertising Practice (BCAP) offers guidance on the interpretation of the UK Code of Broadcast Advertising (the BCAP Code) in relation to broadcast advertisements.
5. The Observer (2023) 'Greenwashing' firms face steep new UK fines for misleading claims. 19 February.
6. See: EasyJet (n.d.) Sustainability. https://corporate.easyjet.com/corporate-responsibility/sustainability
7. Banking on Climate Chaos (2021) Fossil Fuel Finance Report 2021. www.ran.org/wp-content/uploads/2021/03/Banking-on-Climate-Chaos-2021.pdf
8. Ibid.
9. Greenpeace Sweden (2021) A single fine for greenwashing in ten years. 28 September. www.greenpeace.org/sweden/pressmeddelanden/49112/ett-enda-botesstraff-for-greenwashing-pa-tio-ar/
10. Badvertising (2023) Dairy company Arla convicted for misleading green advertising. 1 March. www.badverts.org/latest/dairy-company-arla-convicted-for-misleading-green-advertising
11. George, Sarah (2021) Shell's carbon offsetting ad is greenwashing , rules Dutch watchdog. Euractiv, 2 September. www.euractiv.com/section/all/news/shells-promotion-of-carbon-offsets-is-greenwashing-rules-dutch-watchdog/
12. Reclame Fossielvrij (2022) Shell verliest ook in hoger beroep: CO2-compensatie is misleidend. 21 October. https://verbiedfossielereclame.nl/shell-verliest-ook-in-hoger-beroep-co2-compensatie-is-misleidend/
13. Stratégies (2018) Surconsommation l'Ademe attaque les publicitaires. 2 October. www.strategies.fr/actualites/marques/4018417W/surconsommation-l-ademe-attaque-les-publicitaires.html
14. https://documents-dds-ny.un.org/doc/UNDOC/GEN/N14/499/88/PDF/N1449988.pdf?OpenElement
15. Adfree Cities (2021) ASA report – too close for comfort. https://adfreecities.org.uk/asa/

8 A WORLD WITHOUT ADVERTISING

1. House of Lords Environment and Climate Change Committee (2022) *In our hands: behaviour change for climate and environmental goals.* 1st Report of Session 2022–23. HL Paper 64. https://publications.parliament.uk/pa/ld5803/ldselect/ldenvcl/64/6411.htm#_idTextAnchor126
2. Simms, Andrew with Joe Smith (2008) *Do good lives have to cost the earth?* London: Constable & Robinson.
3. Guardian (2022) Shift your vibe! 60 quick ways to make yourself slightly happier. 22 May. www.theguardian.com/lifeandstyle/2022/may/14/shift-your-vibe-60-ways-make-yourself-slightly-happier
4. Aked, Jody, Nic Marks, Corrina Cordon and Sam Thompson (2008) Five Ways to wellbeing: communicating the evidence. London: nef.

5. Rapid Transition Alliance (2022) Reset series: make public art, not waste. https://rapidtransition.org/resources/make-public-art-not-waste/
6. Simms, Andrew (2023) Hope from the seed of trauma. *New Internationalist magazine*, Issue 542, March–April. https://newint.org/issues/2023/03/01/world-win
7. CAPE (2022) Fossil fuel ads make us sick. https://cape.ca/focus/fossil-fuel-ad-ban/
8. American president: a reference source. Miller Center, University of Virginia.
9. See: www.scenic.org
10. According to a five-year study of 35 cities conducted by the Mississippi Research and Development Center, quoted at: http://www.scenic.org
11. World Without Fossil Ads (2023) Amsterdam 1st city in the world to ban fossil ads. www.worldwithoutfossilads.org/listing/amsterdam-first-city-in-the-world-to-ban-fossil-ads/
12. World Without Fossil Ads (2023) Haarlem bans fossil ads and ads for meat from industrial agriculture. www.worldwithoutfossilads.org/listing/haarlem-bans-fossil-ads-and-ads-for-meat-from-industrial-agriculture/
13. World Without Fossil Ads (2023) Nijmegen adopts motion to ban fossil and industrial agriculture product ads. www.worldwithoutfossilads.org/listing/nijmegen-adopts-motion-to-ban-fossil-and-industrial-agriculture-product-ads/
14. Ibid.
15. Ibid. The Hague looks for ways to ban fossil ads. www.worldwithoutfossilads.org/listing/partial-win-the-hague-looks-for-ways-to-ban-fossil-ads/
16. World Without Fossil Ads (2023). Sydney bans fossil ads and sponsorships. www.worldwithoutfossilads.org/listing/win-sydney-bans-fossil-ads-and-sponsorships/
17. All at: World Without Fossil Ads (2023).
18. Badvertising (2022) Stockholm aims for a fossil ad ban. www.badverts.org/latest/stockholm-aims-for-a-fossil-ad-ban, and Gothenburg, the next Swedish city to ban fossil advertising? www.badverts.org/latest/gothenburg-the-next-swedish-city-to-ban-fossil-advertising
19. World Without Fossil Ads (2023) Lund to ban fossil ads. www.worldwithoutfossilads.org/listing/motion-to-ban-fossil-ads-in-lund/
20. Ibid. Council initiative Turku to ban fossil ads. www.worldwithoutfossilads.org/listing/council-initiative-to-ban-fossil-ads-in-turku/
21. Badvertising (2021) Norwich City Council calls for ethical advertising policy. www.badverts.org/latest/norwich-city-council-calls-for-ethical-advertising-policy
22. Badvertising (2021) Momentum in Liverpool to ban high-carbon ads. www.badverts.org/latest/momentum-in-liverpool-to-ban-high-carbon-ads
23. Badvertising (2021) North Somerset Council votes to end adverts for high carbon products. www.badverts.org/latest/north-somerset-council-votes-to-end-adverts-for-high-carbon-products

24. Cambridgeshire County Council (2022) Cambridgeshire County Council Advertising and Sponsorship Policy. www.cambridgeshire.gov.uk/asset-library/advertising-and-sponsorship-policy-2022.pdf
25. Wald, Richard KC (2023) Legal advice to councils on policies to control high carbon advertising, 39 Essex Chambers, London. Commissioned by the New Weather Institute.
26. Badvertising (2021) Arts against adverts: Bristol's Celebration of Community arts projects. www.badverts.org/latest/arts-against-adverts-bristols-celebration-of-community-arts-projects
27. Adblock Bristol (2017) Local history: St Werburghs billboard campaign. http://adblockbristol.org.uk/2017/04/st-werburghs-billboard-campaign-history/
28. Adfree Cities (2022) 40 monstrous digital screens blocked in 5 years! https://adfreecities.org.uk/2022/08/40-monstrous-digital-screens-blocked-in-5-years/
29. Badvertising (2022) Geneva, the next city to ban harmful advertising? www.badverts.org/latest/geneva-the-next-city-to-ban-harmful-advertising
30. France24 (2014) French city Grenoble bans advertising in favour of trees. 24 November. www.france24.com/en/20141124-french-city-grenoble-bans-advertising-favour-trees
31. Nash, Ogden (1932) Song of the open road. *New Yorker*, 15 October.
32. See: https://zapgames.net/#actions
33. Leow, Jason (2007) Beijing mystery: what is happening to the billboards? *Wall Street Journal*, 25 June. www.wsj.com/articles/SB118273311880146640
34. Guardian (2015) Tehran swaps 'death to America' billboards for Picasso and Matisse. 7 May. www.theguardian.com/world/2015/may/07/tehran-swaps-death-to-america-billboards-picasso-matisse-hockney-iran
35. Libération (2018) Est-il exact, comme l'affirme Nicolas de Tavernost, que les publicités ont été supprimées des chaines publiques britanniques et espagnoles? 13 February. www.liberation.fr/checknews/2018/02/13/est-il-exact-comme-l-affirme-nicolas-de-tavernost-que-les-publi cites-ont-ete-supprimees-des-chaines-_1653161/
36. Congressional Budget Office (2018) *Options for reducing the deficit: 2019–2028*, December, 273.
37. Subvertisers International (2022) Advertising and its discontents. https://subvertisers-international.net/wp-content/uploads/2022/07/Advertising-and-its-Discontents-S.I.-2022.pdf
38. Reporters Without Borders (2023) 2023 World Press Freedom Index. https://rsf.org/en/index
39. Simms, Andrew (2013) *Cancel the apocalypse: the new path to prosperity*. London: Little Brown.
40. Harvard Business Review (2013) You make better decisions if you 'see' your senior self. https://hbr.org/2013/06/you-make-better-decisions-if-you-see-your-senior-self

Index

For topics related to advertising, *see* the topic, e.g. children; colour; tobacco
ill refers to an illustration; *t* to a table

The Pluto Press Newsletter

Hello friend of Pluto!

Want to stay on top of the best radical books we publish?

Then sign up to be the first to hear about our new books, as well as special events, podcasts and videos.

You'll also get 50% off your first order with us when you sign up.

Come and join us!

Go to bit.ly/PlutoNewsletter

Thanks to our Patreon subscriber:

Ciaran Kane

Who has shown generosity and comradeship in support of our publishing.